Radiology of
Small Animal Fracture
Management

Radiology of Small Animal Fracture Management

JOE P. MORGAN, DVM, VET med. dr.
University of California, Davis
Department of Surgical and Radiological Sciences
School of Veterinary Medicine
Davis, California

ROBERT L. LEIGHTON, VMD
University of California, Davis
Department of Surgical and Radiological Sciences
School of Veterinary Medicine
Davis, California

W. B. SAUNDERS COMPANY
A Division of Harcourt Brace and Company

Philadelphia   London   Toronto
Montreal   Sydney   Tokyo

W. B. SAUNDERS COMPANY

A Division of Harcourt Brace & Company

The Curtis Center
Independence Square West
Philadelphia, PA 19106

Library of Congress Cataloging-in-Publication Data

Morgan, Joe P.
    Radiology of small animal fracture management / Joe P. Morgan,
Robert L. Leighton.
      p. cm.
    ISBN 0–7216–5455–X
    1. Dogs—Fractures.   2. Cats—Fractures.   3. Veterinary radiology.
I. Leighton, Robert L. II. Title.
SF991.M67 1995
636.089'715—dc20   94–13109

RADIOLOGY OF SMALL ANIMAL FRACTURE MANAGEMENT   ISBN 0–7216–5455–X

Printed in the United States of America

Last digit is the print number: 9   8   7   6   5   4   3   2   1

We dedicate this book to a gentleman, scholar and scientist who represents the perfect synthesis of both the fields of radiology and orthopedic surgery.

Professor Sten-Erik Olsson

A graduate of the Royal Veterinary College in Stockholm, Sweden, in 1947, he achieved his doctorate from that institution in 1951. Subsequently, he graduated from Karolinska Institutet in 1962 with a medical degree and became docent in experimental surgery at Karolinska Institutet in 1963. He was appointed senior scientist at the Hospital for Special Surgery and Professor of Comparative Orthopedics and Radiology at Cornell University Medical College in New York from 1965 through 1968. He holds an honorary doctorate degree from Guelph University, is an honorary member of the American College of Veterinary Surgeons and the American College of Veterinary Radiology, and holds many meritorious awards including the Bourgelat and Rickard Völker medals.

His most outstanding award is his world-wide reputation with special remembrance by his students, research associates, and colleagues as a friend, scientist, and teacher.

Preface

Management of fractures is both an art and a science. Its development was dependent on the discovery of x-rays, the learning of aseptic techniques, and development of anesthesia. With the expansion of these three specialities, today's surgeon is able to perform feats of skill unknown several decades ago. It is amazing how we have progressed from the use of an inadequately applied external splint to a position today in which the anatomical reduction of a fracture with near-perfect bone healing has become an expected result. Treatment of fractures requires the combination of a knowledge of surgical anatomy, radiographic anatomy, the mechanics of application of either external or internal fixators, and the production of diagnostic radiographs. However, mere knowledge of these specialities is not sufficient. The development of good judgement in fracture management is dependent on the accurate integration of these specialities.

We owe so much to the early pioneers who made crucial contributions to the art and science of veterinary orthopedic surgery and radiology. Memorable advances in small animal surgery were made by E. B. Dibbell, Otto Stader, E. A. Ehmer, Ellis P. Leonard, Erwin F. Schroeder, and Jaques Jenny to name but a few. In radiology we must honor Richard Eberlein, A. Pommer, Mark A. Emmerson, and Gerry B. Schnelle plus many more.

This book is intended to give a real-life approach to interpretation of radiographs of fractures in small animals. From them we plan to extract the most information and to gain as much insight as possible about the best treatment and the most realistic prognosis. By providing an actual series of radiographs of the results of the orthopedic treatment, it enables the reader to view and compare the results of others with similar fractures at hand. It illustrates that success can occur with seemingly inadequate treatment and failure can result even with the best techniques. However, by and large, the odds for success are greatest when all rules of good fracture treatment are observed. The reader is encouraged to make a detailed study of the radiographs, rather than taking only a casual look before proceding on a preconceived course of action. Following evaluation of the radiographs, an appropriate surgical protocol should then be designed that will lead to the ultimate goal of surgery; that is, the return of the patient to normal function as quickly as is possible.

The aim of the text is to make a study of diagnosis and treatment of fractures that permits the clinician in practice or the student in training to more clearly visualize how to achieve a "good" reduction and stabilization followed by bone healing as it "should" occur. In addition they will observe patients in which there was a failure of reduction or stabilization. Both situations are educational as we progress in the study of fracture management.

It is hoped that the book will serve as a supplement to the surgeon's own case records, which may not be as complete or cover as many varieties of injuries.

The cases selected for inclusion within this monograph were chosen from the archives of the Small Animal Radiology Service in the Veterinary Medical Teaching Hospital of the School of Veterinary Medicine at the University of California, Davis, from the years 1970 to 1992. By selecting patients over this span of years, it was possible to include different modalities of fracture treatment with varying results. Patients were treated by both faculty and residents and were not uniformly complimentary. Some cases ending in failure were selected along with those that reached a successful conclusion. It was only in this manner that we felt that we could present the broad picture of fracture management in a more honest manner. The valuable assistance of the many faculty and residents, both of radiology and surgery, is recognized. Their efforts in the collection and usage of this unique material is much appreciated.

During the preparation of this monograph, we recognized the need to discuss the specialities of both diagnostic radiology and orthopedic surgery. The use of radiology requires knowledge of radiography, including the production of a diagnostic radiograph as well as the art and science of their evaluation. In orthopedic surgery, an understanding of the science of fracture management as well as the technical aspects of fragment fixation and stabilization is required. Thus, both the art and science of these specialities need to be blended to achieve the greatest benefit to our patients. Failure in either aspect of radiology or surgery can lead to a less desirable result with disaster lurking around the corner.

While the book includes the discussion of the more common fracture types, it is recognized that not every fracture presented in practice has been included. Some of those not illustrated are uncommon while others require a degree of expertise in treatment not possessed by the average surgeon, and thus, their inclusion was thought to be of a lesser value. We hope that the reader finds that those examples not included are of a limited number and do not seriously hamper the value of the monograph.

J.P.M.
R.L.L.

Contents

x   Contents

Glossary

Acetabular Plate A small bone plate developed especially for positioning over the roof of the acetabulum with both a craniocaudal as well as a mediolateral curvature (also *veterinary acetabular plate*)

Alignment Term used to describe the angle created by the apposing ends of the fracture fragments either before or after reduction

Ankle Mortice The rectangular cavity (created by the malleoli) into which fits the tenon (tarsal bones), creating the tibiotarsal joint (also *hock mortice*)

Anteversion Rotation of the head and neck of the femur in a more cranial direction than the normal 25°–30°

Apical Making reference to the pointed extremity of a conical structure, often used to describe the apex of a tooth and the periapical region

Apophyseal Fracture A type of fracture with separation of a bony apophysis from its parent bone

Apophysis A secondary growth center within a bone that usually provides the site for an important tendinous attachment

Apposition The relationship between the ends of fracture fragments

Arthritis Inflammatory joint disease, usually infectious in nature (also *infectious arthritis, septic arthritis*)

Arthrodesis A surgical procedure intended to lead to bony ankylosis of a joint

Arthrosis Joint Pathology in general—used specifically to refer to noninflammatory joint disease, often after trauma (also *degenerative joint disease, osteoarthrosis, secondary joint disease*)

Articular Fracture A type of fracture, part of which passes through the subchondral bone and articular cartilage (also *intra-articular fracture*)

Articulation Junction between two or more bones

Aseptic necrosis Death of bone due to loss of blood supply

Avulsion Fracture An indirect fracture caused by a ligament or tendon displacing its insertion on a bone

Backed-out Refers to an altered location of an intramedullary pin or screw in which they have shifted in position out of the bone

Bayonneting The placement of the end of a major fracture fragment so that the pointed cortex of one fragment is impacted into the medullary cavity of the other fragment—usually the distal fragment is bayonneted into the proximal fragment

Bone Atrophy Loss of bone tissue with modeling of the bone organ, usually due to disuse but also occurring as a result of loss of vascularity

Bone Callus Trabeculae of woven bone with islands of cartilage that form at the fracture site; visible radiographically because of mineral salts within the osteoid (also *bridging callus, callus*)

Bone Loss The amount of bone tissue lost in the traumatic event

Bone Lysis Loss of bone tissue seen histologically or radiographically and usually associated with an inflammatory process, but also caused by pin motion (also *osteoclastic activity, osteolysis*)

Bone Necrosis Death of bone, often due to thrombosis or thermal event

Bone Plate A stainless steel plate with screw holes used to reduce and stabilize fracture fragments

Bone Screw A stainless steel screw used in orthopedic surgery

Bridging Callus A fibrocartilagino-osseous splint that bridges between major fracture fragments in an effort to recreate the original morphology of the bone (also *bone callus, callus*)

Broad Dynamic Compression Plate A wide stainless steel bone plate with staggered screw holes—the holds provide the opportunity for dynamic compression through screw placement

Bucket-handle Callus A callus that forms in an immature patient with a marked periosteal tear and follows the torn periosteum, which is away from the bone, giving rise to the characteristic pattern of bone formation

Butterfly Fragment A fragment of cortical bone frequently associated with midshaft fractures of long bones that often are amenable to a lagging technique during reduction (also *comminuted fracture, fragmented*)

Buttress Plate Use of a plate to span a highly comminuted section of a diaphysis—a misnomer because the term refers to the manner in which the plate is used, not to a specific plate

Callus A fibrocartilagino-osseous splint that tends to immobilize the bone fragments within a fracture (also *bone callus, bridging callus*)

Calcar Plate of cortical bone on the medial aspect of the femoral neck that supports a femoral head prosthesis or presents an undesirable bony spur after femoral head ostectomy

Cancellous Bone Bone tissue found in the metaphyseal region of long bones, with large marrow spaces between broad trabecular surfaces covered with numerous osteogenic cells

Cancellous Graft Cancellous bone usually harvested from the proximal humerus, ilium, or proximal tibia that, when placed at a fracture site, stimulates fracture healing because of the presence of osteogenic cells

Cancellous Screw A stainless steel screw having a wide thread diameter and a large area between threads (thread pitch)

Cast A molded tubular structure used to provide a degree of stabilization for a fracture

Centerface Pin A type of pin with a positive thread in the center of the pin

Cerclage Wire A stainless steel wire used to encircle the bone and stabilize separated bone fragments or stabilize a long oblique fracture—a misnomer because the term refers to the manner in which the wire is used, not to a specific wire

Chip Fracture A small fracture fragment originating from the corner of a bone, often with an articular fracture line

Chisel Tip The wedge-like shape of the tip of a pin

Cloaca The opening within the bony involucrum that permits drainage from the enclosed sequestrum

Closed Fracture A fracture in which the skin overlying the fracture has not been lacerated or penetrated

Closed Reduction The reduction of a fracture or luxation without surgical incision of the skin

Coapt To bring together the ends of a fractured bone

Coaptation Splint Any splint or supporting bandage that is intended to maintain apposition of the ends of a fractured bone

Comminuted Fracture A type of fracture with many small, interposed bony fragments, often caused by a high-energy injury (also *fragmented, butterfly fragment*)

Compact Bone Bone tissue composed largely of dense bone with few vascular channels, characterized by osteonal formation (also *cortical bone*)

Complete Fracture A type of fracture in which the fracture line extends through the entire cortex with complete separation of the major fragments

Compressed Fracture A type of incomplete fracture in which compression of the fracture fragments occurs, often with loss of bone length (also *impaction fracture, compression fracture*)

Compression The tight fit between fracture fragments after reduction by screws or plating

Compression Fracture A type of fracture in which the fragments are impacted or compressed, often with loss of bone length

Compression Plate Use of a plate in a manner that causes compression of the fragments at the fracture site, or a specific type of plate that can place fragments under compression

Connecting Bars A part of an external fixation apparatus parallel to the bone, attached by clamps to the fixation pins

Connecting Clamps A part of an external fixation apparatus used to attach the fixation pins to the connecting bars (also *clamps*)

Contaminated Refers to an open fracture immediately after injury in which there is dirt and debris within the wound (see *infected*)

Contoured Plate A stainless steel bone plate that has been bent to conform to the bone surface to which it is to be applied

Cortical Bone A dense bone characterized by osteonal formation, forms the outer edge of bone organs (also *compact bone*)

Cortical Remodeling Units Activity within bone tissue resulting in a pattern of osteolysis followed by osteoblastic activity that creates a tube or plate of new bone laid down at the site of bone removal

Cortical Screw A stainless steel screw threaded over their entire length with blunt ends and shallow threads that are closely spaced

Cortical Tunneling The result of cutting cone activity within the cortex that leads to radiolucency within the cortex of a fractured bone; often due to disuse

Coxa Valga A deformity in which the angle formed by the axis of the head and neck of the femur and the axis of the shaft of the femur is greater than normal, as measured on the axial surface of the bone

Coxa Vara A deformity in which the angle formed by the axis of the head and neck of the femur and the axis of the shaft of the femur is less than normal, as measured on the axial surface of the bone

Countersink The placement of the end of a pin below the surface of the bone by use of a pin setter

Crepitation The abnormal sound or feeling detected when fractured bones are palpated, caused by movement of the bony fragments against each other (also *crepitus*)

Cross Pinning A technique used in fragment reduction; small wires or pins are placed in such a manner that they cross each other, forming an "X"

Cutting Cone The cellular activity that results in osteoclastic activity within bone—often followed by osteoblastic repair—evident in fracture callus modeling or primary fracture repair

Cut-to-Length Plate A stainless steel bone plate prepared in long lengths that can be cut to a required length for use in treatment of a particular fracture

Cycling Repetitive motion due to instability of the fracture, leading to failure of a pin or plate

Degenerative Joint Disease General joint pathology—specifically, noninflammatory joint disease, often after trauma (see *arthrosis, osteoarthrosis, secondary joint disease*)

Degloving Injury Severe soft tissue injury with removal of skin and subcutaneous tissue after an abrasive type trauma—usually distally on a limb (also *shearing injury*)

Delayed Union A subjective evaluation of fracture healing that is not occurring in an expected manner given the patient, nature of the fracture, and character of the reduction and stabilization (also *delayed healing*)

Depth Gauge A device used to determine the required length of a bone screw

Dislocation A complete displacement of joint components (also *luxation*)

Dorsal Toward the upper surface of all four feet

Drill Holes Holes drilled in bone before reception of a pin, wire, or screw

Dynamic Compression Plate (DCP) A special stainless steel bone plate that achieves compression at the fracture site through the use of special holes which function as inclined planes to bring the fragments together and compress them at tightening of the screws

Elephant Foot Callus A specific form of nonbridging, mature bony callus at a fracture site, characterized by a large, dense, smooth shape (elephant foot) and indicating nonunion

Ellis Pins A type of stainless steel fixation pin with a very short threaded tip behind a trocar point

Epiphysitis Inflammation of the epiphyses, usually due to an infectious agent

Exuberant Callus Large bridging callus; increased size due to motion at the fracture site, requiring a massive callus

External Callus Proliferation of mesenchymal cells arising from the osteogenic layer of the periosteum and paraskeletal soft tissues, forming the callus around the fractured bone

External Coaptation Use of devices such as casts or splints to stabilize bone fragments through external application to the skin

External Fixators Fixation devices consisting of fixation pins, connecting bars, and clamps (also *external fixation, K-E device, K-E apparatus*)

External Rotation Outward rotation of a limb away from the midline

Extra-articular Fracture A type of fracture without involvement of an articular surface

Extraosseous Blood Supply of Healing Bone The new blood supply originating from surrounding soft tissues that provides the blood supply for the formation of the early external callus

Eyed Cerclage Wire A special form of wire to be used in a cerclage technique with a preformed eye or loop through which the end of the wire is drawn using a special tightener—the end is bent back on itself, securely locking the wire in place

Fatigue Fracture A type of bone fracture caused by repetitive stress—fracture of a bone plate after repetitive stress (also *greenstick fracture, incomplete fracture, march fracture, stress fracture*)

Fibrous Callus The early callus that forms around the fracture; lacks mineral density

Figure-8 Compression Band Wiring Placement of a wire in a figure-8 pattern between two pins, screws, or holes (or a combination thereof) in the bone fragments, in which the tightened wire compresses the intervening bone separation

Figure-8 Wire Use of a stainless steel wire placed in a figure-8 pattern in compression band wiring

Finger Plate A small stainless steel bone plate originally made for plating fractures of the human hand, but adaptable to treatment of many small bone fractures in the dog and cat

Fixation Pins Those parts of an external fixation apparatus that are placed through surrounding soft tissues and bone, and attached to the connecting bars with clamps

Fixator Any internal or external device used to reduce and stabilize a fracture

Fracture An injury to a bone in which the stresses exceed the strength of the bone, and a break(s) occurs

Fracture Healing A subjective evaluation of the status of a fracture made radiographically, through physical examination, or by noting the clinical status of the patient

Fracture–Luxation A special injury in which there is a combination of a fracture and a luxation, with the fracture lines often entering the injured joint

Fragmented A fracture with multiple fragments (also *comminuted fracture, butterfly fragments*)

Full-pin Fixation A type of external fixator with the pins passing through both the medial and lateral skin surfaces

Greenstick Fracture A type of fracture in which a portion of the cortex remains unbroken (also *fatigue fracture, incomplete fracture, march fracture, stress fracture*)

Growth Plate Anatomically the zone between the metaphysis and epiphysis in a skeletally immature animal from which the bone grows in length (also *physeal plate*)

Half-pins A type of small pin with double trocar tips and a smooth end and a threaded end, made for use with acrylic splints in treatment of long bone fractures in toy breeds, or in treatment of mandibular fractures

Half-pin Fixation A type of external fixator with the pins passing through only one skin surface

Healed Fracture A subjective evaluation of the status of a fracture made radiographically, through physical examination, or by noting the clinical status of the patient, suggesting that the patient can use the limb safely for at least limited physical activity without fear of refracture

Hemarthrosis Hemorrhage within a joint

Hematogenous Osteomyelitis A form of bone infection in which the infectious agent arrives at the bone through the blood supply

Hemicerclage Wiring The use of stainless steel wire to encircle the bone partially and stabilize separated bone fragments or stabilize a short oblique fracture—the term refers to the manner in which the wire is used, not to a specific type of wire

High-energy Injury A type of trauma resulting in severe soft tissue injury with comminution at the fracture site

Hock Mortice The rectangular cavity (created by the malleoli) into which fits the tenon (tarsal bones), creating the tibiotarsal joint (also *ankle mortice*)

Homeostasis The tendency toward stability in the internal environment of the normal body

Iatrogenic Disease or injury occurring as a result of the surgeon's intervention

Impaction A type of incomplete fracture in which compression of the fracture fragments occurs, often with loss of bone length (also *impacted fracture, compression fracture*)

Incomplete Fracture A type of fracture in which a portion of the cortex remains unbroken (also *greenstick fracture, fatigue fracture, march fracture, stress fracture*)

Infected Refers to an open fracture several hours (often 8 hours) after injury in which dirt and debris within the wound have resulted in infection (see *contaminated*)

Infectious Arthritis Inflammatory joint disease infectious in nature (also *arthritis, septic arthritis*)

Indirect Fracture A type of fracture occurring at a distance from the point of trauma

Interfragmentary Compression A technique of fracture reduction that places the fragments under compression through the use of a lag screw technique

Internal Callus The proliferation of mesenchymal cells arising from the osteogenic cells that move into the fracture site, resulting in later mature callus formation (also *medullary callus*)

Internal Fixation Techniques of fracture stabilization using implants contained entirely beneath the skin

Internal Rotation Inward rotation of a limb toward the midline

Intra-articular Fracture A type of fracture, part of which passes through the subchondral bone and articular cartilage (also *articular fracture*)

Intramedullary Fixation A type of fracture stabilization in which implants are placed within the medullary cavity of long bones (commonly *IM*)

Involucrum Pathologically and radiographically, the dense bone that forms around a sequestrum, creating a kind of cavity wall

Irritation Callus Callus forming at a fracture site without the pattern of bridging stimulated by motion at the fracture site

K Wires Small pins used to stabilize bone fragments (also *Kirschner wires*)

K-E Apparatus A fixation device consisting of fixation pins, connecting bars, and clamps (also *external fixator, K-E device, Kirschner-Ehmer apparatus, Kirschner-Ehmer device*)

Kirschner Wires Small stainless steel pins used to stabilize bone fragments (also *K wires*)

Küntscher Nail A stainless steel pin shaped as a "V" or cloverleaf used in intramedullary fixation

Kyphosis A dorsal curvature of the spine as seen from the side

Lag Screw The particular use of a screw in obtaining interfragmentary compression in such a manner that the far fragment is drawn tightly against the near fragment—a misnomer because the name describes the technique, not a particular screw (also *lag technique*)

Lordosis A ventral curvature of the spine as seen from the side

Low-energy Fracture A type of fracture sustained after minimal trauma in which soft tissue injury is slight and the fracture is characterized as simple without comminution and/or fragment displacement

Lubra Plate (<u>Lum</u>b and <u>Bras</u>mer) A longitudinally curved plastic plate originally devised to be bilaterally applied to the dorsal vertebral spines in treatment of spinal fractures; sometimes used in a tension-band fashion for dorsal acetabular fractures

Luxation Complete displacement of joint components (also *dislocation*)

Malunion A variation from desired fracture healing with malalignment, angulation, or rotation of the major fracture fragments

March Fracture A type of fracture caused by repetitive stress (also *fatigue fracture, greenstick fracture, incomplete fracture, stress fracture*)

Mason Metasplint A semitubular, aluminum lower forelimb splint with a cup-shaped, flared end to accommodate the foot—additional sections can be attached to provide extra length (*meta* = metacarpal)

Medullary Callus The proliferation of mesenchymal cells arising from osteogenic cells that move into the fracture site and cause later, mature callus formation (also *internal callus*)

Methylmethacrylate A quick-hardening plastic used to assist in fracture stabilization

Micromotion Minimal but repetitive motion at a fracture site that leads to delayed union, nonunion, or failure of a fixator

Miniplate A miniaturized version of regular bone plates devised for cats and small and toy breeds of dogs (also *toy plates*)

Modified Rush Technique A modification of the original Rush technique that was developed for humans, consisting of placement of two thin, flexible IM pins with a chisel point on one end and a hooked tip on the other end, to stabilize distal or proximal long bone fractures; proper application results in three-point contact on the endosteal surfaces and very stable fixation—the modification includes use of pins without the hooked tip

Modified Thomas Splint A modification of the original Thomas splint that was developed for humans, an aluminum rod device for the treatment of fractures within the distal portions of the forelimbs and hindlimbs, consisting of a padded circular portion that fits tightly into the axilla or groin with cranial and caudal bars that extend distally to a "U"-shaped piece distal to the foot—variations in the bends in the bars and the degree of bandaging provide for traction and coaptive control of fracture fragment position (also *Schroeder-Thomas splint*)

Monteggia's Fracture A fracture in the proximal half of the ulna combined with luxation of the radial head

Mortice The rectangular cavity (created by the malleoli) into which fits the tenon (tarsal bones), creating the tibiotarsal joint (also *ankle mortice, hock mortice*)

Neocortex The cortex resulting from modeling of the bone tissue at the fracture site that tends to resemble the old cortex

Neutralization Plate Use of a stainless steel plate to protect a fracture characterized by bone loss from excessive weight-bearing stresses—a misnomer because the term refers to the manner in which the plate is used, not to a specific plate

Nonunion A fracture in which all osteogenic activity has ceased without bony union of the fragments, as determined radiographically

Normograde Placement of a pin within the medullary cavity by starting the pin at the proximal end of the bone and directing the pin distally

Normoversion Rotation of the head and neck of the femur in the normal direction of 25°–30° anteversion

Nutritional Secondary Hyperparathyroidism A disease leading to severe osteopenia with pathologic fractures

Oblique Fracture A type of fracture in which the fracture line extends at an angle compared with a line drawn transversely across the bone

Open Fracture A type of fracture in which there is an external wound

Open Reduction The reduction and stabilization of a fracture or luxation using surgical exploration

Orthogonal Views Two radiographic studies made at right angles to each other

Orthopedic Historically, the straightening of the congenitally deformed foot of a child—currently, the preservation and restoration of function of the skeletal system and associated structures (Greek *ortho* = straight, *pais* = child)

Orthopedic Wire A biocompatible stainless steel wire that can remain within the body indefinitely, made to withstand the stresses of tension and twisting when used in cerclage and figure-8 formations

Ostectomy The surgical removal of a piece of bone (also *ostectomy site*)

Osteoarthrosis General joint pathology—specifically, noninflammatory joint disease, often after trauma (also *arthrosis, degenerative joint disease, secondary joint disease*)

Osteoconduction Surgical technique consisting of the use of a cancellous graft to provide a scaffold on which callus formation can take place

Osteoinduction The recruitment of mesenchymal or pleuripotential cells from a cancellous graft in callus formation, as seen histologically

Osteoclastic Activity Loss of bone tissue as identified histologically or radiographically, and usually associated with an inflammatory process, but also caused by pin motion (also *bone lysis, osteolysis*)

Osteolysis Loss of bone tissue as identified histologically or radiographically, and usually associated with an inflammatory process, but also caused by pin motion (also *bone lysis, osteoclastic activity*)

Osteomalacia Inadequate mineralization with normal osteoid

Osteomyelitis General inflammatory condition of a bone associated with infection

Osteopenia Any general decrease in bone mass to a level below that of normal

Osteoperiostitis A nonspecific infection within the periosteum spreading to the underlying cortical bone

Osteoporosis Loss of bone mineral and osteoid to a level below that of normal

Osteophytes Dense, sharply defined bony proliferations, as seen radiographically or surgically

Osteosynthesis The surgical fastening of the ends of a fractured bone by sutures, plates, or other mechanical means

Osteotomy The technique of surgical cutting of a bone

Osteotomy Fragment The fragment of bone separated by the use of an osteotomy technique

Osteotomy Replacement Pins Small pins or wires used to replace a bone fragment cut surgically to provide surgical access to a region

Palmar The surface of the foot of the thoracic limb that is in contact with the ground

Panarthrodesis The surgical procedure leading to bony ankylosis of all joints

Pathologic Fracture A type of fracture occurring through earlier diseased and weakened bone

Penciling Descriptive of bone atrophy due to disuse in which the ends of the bony fragments taper in a pencil shape

Periosteal New Bone The sheet-like reactive bone originating from the osteogenic layer of the periosteum

Physeal Fracture A type of fracture involving all or a part of a physeal plate

Physeal Plate Anatomically, the zone between the metaphysis and epiphysis in a skeletally immature animal from which the bone grows in length (also *growth plate*)

Physiotherapy The use of physical therapy— active or passive manipulation of a limb—to maintain soft tissue health as well as the health of the bone during fracture healing

Physitis Inflammation in the growth plate, usually due to an infectious agent

Pin–Bone Interface The site of weakness within a pin at the point where the pin enters the cortex

Plantar The surface of the foot of the pelvic limb that is in contact with the ground

Plate Removal The time or procedure used in removal of a bone plate, ideally after healing of the underlying fracture

Positive Thread A type of pin in which the thread diameter is slightly larger than the shank diameter

Prestressing Slight bending of the plate to create a small gap under the plate, resulting in the fracture site being placed under compression and the plate being placed under tension when the screws positioning the plate are tightened

Pronate Placement of the foot with the dorsal surface uppermost

Pseudarthrosis The pathologic entity resulting from a nonunion fracture with cartilage forming over the ends of the bone, creating a "false joint"

Quadriceps Tie-down A severe complication of femoral fracture repair characterized by hyperextension of the stifle and hock joint

Reconstruction Plate A specialized bone plate made to be conformable in two planes of space by means of angled indentations along its length

Reduction The technique of correction or re-alignment of fracture fragments, or replacement of a luxated articular component

Retrograde Placement of a pin within the medullary cavity by starting the pin at the distal end of the bone and directing the pin proximally

Retroversion Rotation of the head and neck of the femur in a more caudal direction than the normal 25°–30° anteversion.

Robert Jones Bandage A heavily padded bandage applied to a limb with a major fracture, used to prevent further damage to the soft tissues and to minimize swelling

Rotation Describes the relationship between the ends of the fracture fragments relative to turning around the long axis

Rush Pins A flexible type of intramedullary pin used in pairs with a chisel point on one end and a hooked tip on the other end, achieving three-point contact on the endosteal surfaces (also *Rush pin technique*)

Salter-Harris Fracture Classification A system for categorizing fracture types that involve the physeal plates of long bones

Schroeder-Thomas Splint A modification of the original Thomas splint that was developed for humans, an aluminum rod device for the treatment of fractures within the distal portions of the forelimbs and hindlimbs, consisting of a padded circular portion that fits tightly into the axilla or groin with cranial and caudal bars that extend distally to a "U"-shaped piece distal to the foot—variations in the bends in the bars and the degree of bandaging provide for traction and coaptive control of fracture fragment position (also *modified Thomas splint*)

Scoliosis Lateral curvature of the normally straight line of the spine

Screw-tipped Pin A stainless steel pin with a threaded segment on the pointed end

Seated Refers to the location of the end of the pin within the metaphysis/epiphysis, suggesting the firmness of the position (also *deeply seated*)

Secondary Joint Disease Specifically, noninflammatory joint disease, often after trauma (see *arthrosis, degenerative joint disease, osteoarthrosis*)

Segmental Fracture A type of fracture in which an intermediate fragment with a complete cortex is created between two shaft fractures

Semitubular Bone Plate A specialized bone plate that achieves lightness and strength because of its tubular structure

Septic Arthritis Inflammatory joint disease, infectious in nature (also *arthritis, infectious arthritis*)

Sequestrum Nonviable bone created through loss of blood supply, located within an infected soft tissue environment

Sequential Radiographs A radiographic study made during the course of fracture identification, repair, and healing with the intent of profiting by comparison with an earlier study

Shearing Injury Severe, abrasive injury that removes skin, subcutaneous tissue, and various amounts of bone (also *degloving injury*)

Sinus An abnormal channel permitting the escape of pus, such as from a sequestrum

Small Fragment Plate Very small bone plates specifically devised for application in the reduction of small fragments

Smooth Pin A type of stainless steel pin without threads

Spiral Fracture A type of fracture in which the fracture line is a long spiral often occupying most of the shaft of the bone

Sprain Injury to a ligament with rupture of some of the fibers resulting in joint injury

Stabilization A subjective evaluation of the nature of the fracture repair, suggesting that it will withstand all forces during healing and ultimately heal successfully (also *stable fracture*)

Stacked Pins The technique of using more than one intramedullary pin to achieve greater stability at the fracture site

Standard Thread Profile A type of stainless steel pin with a thread that has the same diameter as the shank diameter

Step The uneven nature of the fracture surfaces, creating the possibility of stability after reduction as a result of the tight fitting together of the fragments

Strain Less severe than a sprain; injury to a muscle or tendon resulting in joint injury

Stress Fracture A type of fracture caused by repetitive stress (also *greenstick fracture, fatigue fracture, incomplete fracture, march fracture*)

Stress Protection Refers to the atrophy of bone associated with a rigidly fixed fracture where the implant absorbs most of the stress

Stress Studies Radiographic studies made with the limb or bone stressed in a medial or lateral, flexed or extended, or rotational position to demonstrate abnormal joint laxity

Steinmann Pin A specific type of pin most commonly used as an intramedullary implant

Subluxation Incomplete displacement of joint components—often a subjective evaluation

Supercondylar Fracture A type of fracture located in the distal end of the humerus or femur in which the fracture line is just proximal to the condyles

Supportive Bandage Any form of external bandaging that limits further soft tissue injury to the limb or additional movement of the fracture fragments

Supinate Placement of the foot with the palmar or plantar surface uppermost

Symphyseal Fracture A type of fracture with separation of the fibrocartilage junction uniting two parts of a bone such as the mandibular or pubic symphysis

Synostosis An abnormal union between adjacent bones or fragments of bones by bony or fibrous callus

"T" Fracture A fracture, most commonly distal humeral, with a configuration characterized by a vertical arm within the middle of the distal humeral condyle and horizontal arms extending laterally and medially just proximal to the condyle

Tension Band An engineering term that refers to a pin and wire device or a bone plate placed under tension during loading (also *tension-band device, tension-band K wires*)

Tension Side The surface of a bone under bending stress that is being pulled apart

Threaded Pin A type of pin with threads along the shaft (*threaded tip, threaded center*)

Threaded Tip The tip of a pin that is pointed and threaded

Toy Plate A miniaturized version of regular bone plates devised for cats and small and toy breeds of dogs (also *miniplates*)

Trabecular Bone Interconnected plates of cancellous bone found within the metaphyses of long bones and beneath articular surfaces

Transcondylar The location of a pin or screw that passes through both distal humeral or femoral condyles, achieving reduction, stabilization, or both

Transverse Fracture A type of fracture in which the fracture line is perpendicular to the long axis of the bone

Trocar Tip The tip of a pin that is pointed and in the form of a three-sided pyramid

Tubular Bone Plate A specialized bone plate that achieves lightness and strength because of its tubular structure

Type I Splintage A type of external fixator with the pins passing through only one skin surface

Type II Splintage A type of external fixator with the pins passing through both medial and lateral skin surfaces

Type III Splintage A type of external fixator combining types I and II techniques

Unstable Fracture A subjective assessment of the nature of the fracture repair, suggesting that it will not withstand forces during healing and will lead to delayed union or nonunion

"Y" Fracture A type of fracture, most commonly distal humeral, with a configuration characterized by a vertical arm within the middle of the distal humeral condyle and two horizontal arms that extend from each cortex and converge medially just proximal to the condyle

Valgus A conformational abnormality in which the angle measured on the axial surface of the bone or limb is greater than normal

Varus A conformational abnormality in which the angle measured on the axial surface of the bone or limb is less than normal

Veterinary Acetabular Plate A special bone plate specifically designed for treatment of canine acetabular fractures—adapted to the craniocaudal and lateral curvature of the bones dorsal to the acetabulum (also *acetabular plate*)

Woven Bone Immature bone first formed at the fracture site, without trabecular pattern

Introduction

In our interest in orthopedic surgery, we must not forget that a fracture is contained in a body all parts of which are equally susceptible to damage, and that the body is responding to the trauma with a variety of physiologic changes resulting from the injury while at the same time endeavoring to maintain homeostasis and initiate injury repair. At first, all immediate efforts of the veterinarian are focused on saving the life of the animal. A patent airway to provide for delivery of oxygen, intravenous catheterization for administration of blood and fluids, and stabilization of effective cardiac function are required. Some temporary splinting such as a modified Robert Jones, a Schroeder-Thomas, or a coaptation splint controls some long bone fractures, preventing further damage to fracture fragments and further soft tissue damage. With the patient stabilized, examination for respiratory, abdominal, and neurologic injury is done. Enthusiasm for treating the obvious fracture should be tempered by the possibility of the presence of concomitant injuries that may require more immediate treatment.

The examination of the total patient is enhanced by radiography. With evidence of respiratory distress, dulled heart sounds, or abnormal chest sounds, thoracic radiographs may reveal a ruptured diaphragm, torn lung lobe, fractured rib, hemothorax, pneumothorax, or collapsed lung lobes, the treatments of which have priority over fracture repair. Among the abdominal injuries seen in trauma patients are crushed liver, damaged kidneys, ruptured bowel, ruptured spleen, ruptured bladder, and torn urethra (Fig. 1-1). The pregnant patient may have a ruptured uterus.

The sometimes neglected neurologic examination is especially necessary after trauma. The possibility of brain and spinal cord injuries as well as peripheral nerve damage should be considered. Radial, femoral, ischiatic, and pudendal nerve injuries may be such that amputation or euthanasia

is required. Often these injuries are difficult to detect because the patient is obtunded and unable to respond to stimuli. Radiographs of the spine can identify areas of disc herniation and vertebral fracture–luxations that often are associated with spinal cord injury.

Only after the clinical examination is completed, is attention given to the orthopedic examination. Gentle palpation and manipulation can provide many clues as to the extent of the damage. Other regions should be carefully examined for injuries. In the complete examination of the patient, orthopedic injuries in addition to those that are obvious may be found. It is not unusual to have a fracture on one side of the patient and a luxation on the contralateral side. Thoracic and pelvic limbs can be simultaneously injured. Spinal fractures that are clinically silent may be found in the patient that is recumbent because of long bone fractures.

The physical findings should be recorded in detail within the clinical record at the time of the original examination. Memory may fail when the record is written up in final form a few days later. The question of whether one did or did not examine a particular part may be critical not only for medical but for legal reasons. The written record made at the time is the only good defense, and is good medical practice.

The radiographic examination provides the most accurate, permanent, and irrefutable evidence of bone and joint injury. Evaluation of skeletal injury requires examination of radiographs made in two planes (Fig. 1-2). Orthogonal views permit the creation of a three-dimensional image of the injured part. The physical findings aid the radiographer in determining the anatomic regions to be examined. It is important that the radiographic studies be made in conjunction with the findings from the physical examination. It is not good medicine to select sites for radiography on

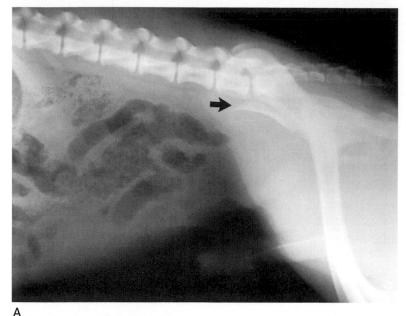

A

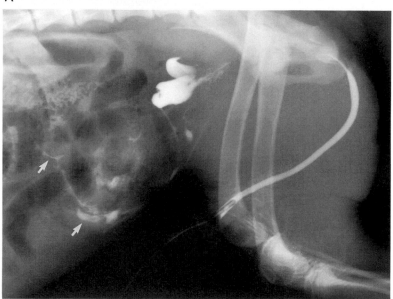

B

FIGURE 1-1. Lateral abdominal radiographs of a male dog after trauma who was not seen to urinate. Pelvic fractures (arrow) are seen on the survey radiograph (A). After retrograde administration of positive contrast agent within the urethra, free contrast agent is seen within the peritoneal cavity (arrows, B). The fracture should not receive the clinician's full attention until the possibility of concomitant soft tissue injuries has been considered.

the basis of the owner's observations without having made a complete physical examination determining the extent of the damage. Radiographic views are labeled in a standard manner using terminology relative to the cranial and caudal surfaces of the limb and the dorsal and palmar–plantar surfaces of the foot. The description of the view indicates the direction of the x-ray beam as it passes through the body; however, in a lateral

view, the x-ray beam is directed from either the medial or lateral side (Fig. 1-3).

The determination of the mode of repair and the prognosis after identification of the fracture depends on a number of considerations in addition to the location and variety of the fracture itself (Manley et al, 1981). The presence and extent of soft tissue injury, contamination or infection, limited blood supply, neurologic injury, and concomi-

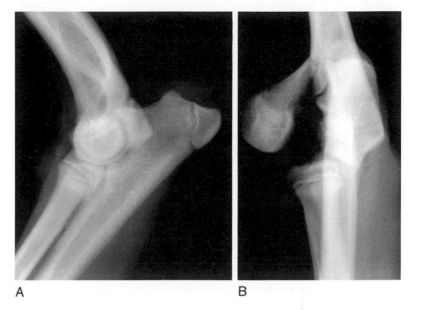

FIGURE 1-2. Radiographs of a puppy with a severe elbow injury. The condylar fracture is not identified on the lateral view (A), although it is easily identified on the craniocaudal view (B). Two radiographic views are required for a complete understanding of skeletal injury.

A   B

tant fractures and luxations alter the choices for repair. The blood supply to a limb may be badly compromised or a massive hematoma may irreparably damage the surrounding musculature. A partially severed foot often may be deprived of a blood supply; a partially avulsed tail may become totally necrotic.

Surgical choices are modified by the presence of preexisting chronic injuries such as those seen secondary to hip dysplasia, elbow dyscrasias, and

osteoarthrosis of the stifle joint. It is unwise to proceed with a reparative procedure that places undue stress on such joints. The radiographic study should be broadened to provide examination of adjacent bones and joints, especially those frequently affected by developmental bone and joint disease.

Several specific disease processes can cause focal or generalized bone disease with resulting pathologic fractures that interdict or alter treatment.

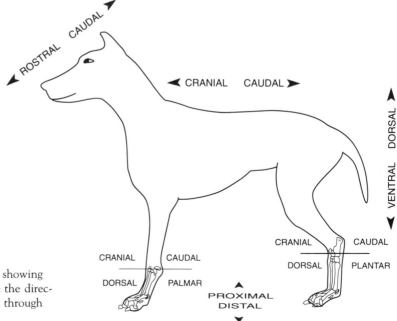

FIGURE 1-3. Drawing of a dog showing the nomenclature used to describe the direction of the x-ray beam as it passes through the body.

These include primary bone tumors, renal osteo-dystrophy, developmental bone lesions, and nutritional bone disturbances. These unusual causes of fractures must be detected and the mode of treatment of the patient altered to take into consideration the actual cause of the fracture.

Often the prognosis of a fracture is based on the owner's ability to assist with postfixation care. The elderly couple may not be able to help their Saint Bernard dog to its feet, or the business couple may not have the time to provide exercise and physiotherapy as required. Physiotherapy strengthens muscles and preserves function and may be a requirement for limb function after fracture healing, but consideration must be given to by whom and how it is to be administered. Nursing care and administration of medication is often the key to good recovery. The time and skill required for this assistance and its importance relative to healing must be discussed with the owner.

With the determination of the diagnosis, the client must be informed as completely as possible of the appropriate mode of treatment and a reasonable expectation as to the outcome. This requires the art of tact, persuasion, instruction, understanding, compassion, psychology, and patience. Ultimately comes the discussion of costs. Proper fracture treatment often is expensive, and the surgeon cannot follow it as a hobby. What is best for the patient should be held uppermost. The selection of therapy based on projected cost alone often is not the best selection either for the patient or for the owner's pocketbook. Often what is thought of as an inexpensive form of treatment becomes progressively more costly as repeated visits are required for adjustment and evaluation of the healing fracture.

The nature of the patient must be understood so that the appropriate form of treatment is se-lected. The owner not only needs to understand the limits of how the surgeon can assist in the injury repair, but how important is the limitation of the patient's activities. Exuberant dogs, let run too soon, can fracture bone plates, tear off external fixation devices, destroy casts, and undo the excellent work of the surgeon.

With these admonitions in mind, let us observe by means of radiographs the fractures, treatments, and results in actual cases. The commentary to follow is never meant to be disparaging, only instructive. Often, healing occurs even when the repair was not performed in strict accordance with recognized techniques, and sometimes repair fails even though the rules of treatment have been followed in a rigid manner. It is through examination of fracture cases that much can be learned about the radiography of fractures, the interpretation of the radiographic findings, and subsequent fracture healing. Subtle messages often are provided on the films to those willing to study them.

Both the *dog and the cat* have provided examples of fractures that are used in this text. Reasons for selection of various modes of treatment often are based on the species as well as on the type of injury, the bone involved, and the expected reaction of the patient to the technique of fracture stabilization.

The function of this book is to describe how radiography can be useful in diagnosis and treatment of the fracture patient. The case histories clearly show how events can proceed as expected, and sometimes in ways not anticipated. The text is written to supplement the excellent descriptive text, *Handbook of Small Animal Orthopedics and Fracture Treatment*, by Brinker, Piermattei, and Flo.

Introduction to Radiography 2

The *radiographic diagnosis* of trauma to the musculoskeletal system is most conveniently broken into divisions because of the great variation in radiographic appearance of traumatic lesions. This is the result of the differences in the character of the bony structures and the differences in the radiographic techniques that are used in examination. The divisions are those containing: (1) the long bones, (2) the interposed joints, (3) the spinal column and interposed disc spaces, and (4) the head (Morgan, 1972). Injury to diarthrodial joints in the limbs may accompany long bone fractures or may present as a separate entity. Evaluation of joint injury is considered separately because of the difference in radiographic changes, the possibility of using stress radiography, and because of the difference in the clinical significance of injury to the joints. Although radiographic evaluation of the soft tissues associated with these bone and joint lesions is limited, some comments are included that may provide additional information in trauma cases.

Sequential radiographs are made during the course of a fracture and its repair. The first study determines the presence and character of the injury. In addition, these often show subtle findings that influence the selection of mode of treatment and the prognosis of fracture healing. Next, the postreduction or poststabilization radiographs provide information as to the character of injury healing and the expected time frame in which the healing will occur. Prognosis as to the athletic ability of the patient after injury can be predicted from these posttreatment studies. Additional radiographs often are made during the healing stages in the event of the development of clinical signs suggesting that healing is not proceeding as expected. The accurate evaluation of the posttreatment radiographic studies provides information as to healing that may be easily overlooked. The radiographic appearance of the healing fracture is referred to frequently, and its significance considered (Fig. 2-1).

It is not possible to discuss specific *radiographic techniques* for each of the fracture cases discussed. Patient positioning and machine settings as well as type of radiographic film and intensifying screen can be studied in other textbooks (Morgan, 1993). Particular problems of patient positioning relative to a specific fracture or special problems in obtaining a diagnostic radiograph will, however, receive some comment. It is important to remember that it is not possible to gain complete information concerning the injury unless a complete radiographic study is available for evaluation. Because of their tubular shape, long bones are more easily evaluated radiographically than the joints or the axial skeleton. It is essential in the development of a plan for radiographic views that the anatomic region be considered, as well as particular problems in positioning that might be present in a given patient because of pain or temporary splinting of the limb. The use of a horizontal x-ray beam provides a method of positioning the patient that is often less painful (Fig. 2-2). The use of sedation or anesthesia greatly influences how the radiographic views can be made. Selection of the radiographic technique usually is limited to two orthogonal views; however, the careful evaluation of these views to determine the nature of fracture healing is crucial if an infected fracture or a potential nonunion or malunion fracture is to be recognized as early as possible. The early determination of a problem is important so that a secondary method of injury treatment can be considered before soft tissue atrophy and lack of osteoblastic potential becomes so extensive that bone healing is difficult to obtain.

Radiographs must be readily available for *review* and must be carefully *identified*. This is especially important because radiographs are made of a fracture patient on different dates. Storage of the

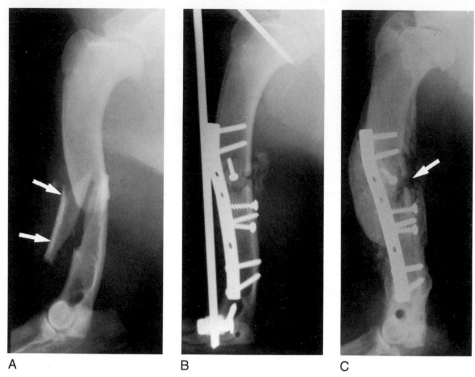

A   B   C

FIGURE 2-1. Sequential radiographs of the left humerus in a 1-year-old female Great Pyreenes that was struck by a speeding car. Only the lateral views are included. The first radiograph (A) serves to identify the nature of the fracture, with special attention called to the large butterfly fragment (arrows). The postreduction radiograph (B) is used to evaluate the placement of interfragmentary screws, the bone plate, and the external fixator. The progress study (C) was made at 50 days posttreatment after removal of the external device, and showed failure of callus to bridge the fracture site caudally (arrow). This series of radiographs demonstrates the manner in which radiography can be used to assist in fracture management.

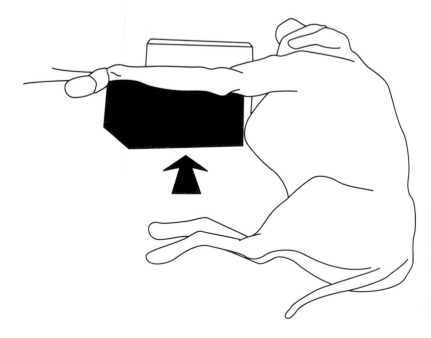

FIGURE 2-2. Drawing of a dog in lateral recumbency illustrating the use of a horizontally directed x-ray beam in a technique for radiography of the radius and ulna that is often less painful to the patient.

radiographs must permit the expeditious recovery of films as required. Unfortunately, in a litigious society, it also is important to be able to show evidence of the examinations and treatment the patient received.

The higher the level of *client knowledge* about their pet, the better informed they are relative to the injury and its repair. One of the best methods of informing clients of their pet's injury and expected recovery is through the use of radiographs. Obviously, the higher the quality of the radio-graphs, the easier it is to use them as a method of client education. Radiographs obviously play a major role in conferences that review the handling of orthopedic cases within the hospital.

Radiation safety is important to those assisting in making radiographic examinations. Routine techniques of safety can be reviewed in general radiology textbooks. Special comments are included in this book only if a unique situation is presented because of the character of the patient injury.

Fracture Description

3

Fracture Diagnosis

Fracture diagnosis is most easily done by the identification of a break in the cortical shadow, usually seen on the radiograph as a radiolucent zone. In bones that do not have a strong cortical shadow, such as the maxilla, fracture identification is more difficult. Fractures in immature bones or osteopenic bones in the mature patient also are more difficult to see because the underlying cortical shadow is not as prominent radiographically, and the cortical disruption often is much less than that seen in fractures of healthy bones. Obviously, the amount of separation of the fracture fragments greatly influences the ease of identification of the lesion on the radiograph. The greater the separation of the bony fragments, the easier it is to identify the bone injury.

Fracture Classification

Fracture classification can be determined radiographically through examination of the character of the underlying bone (Morgan, 1978) (Table 3-1). A *direct injury to healthy bone* causes most of the fractures presented to the clinician. Pathologic fractures occur, however, and the underlying diseased bones must be recognized on the radiograph to avoid treatment of the fracture. *Pathologic fractures* occur most commonly in patients with nutritional secondary hyperparathyroidism. This generalized bone disease is seen less often today because most owners feed a commercially prepared food that has appropriate levels of calcium and phosphorus. Cats with this type of bone disease are most commonly affected today owing to the unique eating habits displayed by both owners and patients. Pathologic fractures also occur through bone weakened by a malignant process. If there is only minimal cortical thinning, identification of

the pathologic nature of the fracture may be difficult (Fig. 3-1).

Indirect injury to bone may occur when excessive loads are transmitted through a bone to a more distant location. A tibial tuberosity avulsion may occur from quadriceps contraction. *Stress or fatigue fractures* occur from repeated stress on the bone causing microdamage. Ultimately, the repeated nature of the stress results in an incomplete fracture.

Fractures may be classified depending on the extent of damage to the bone. In *complete fractures*, the fracture line extends through the entire cortex, with complete separation of the fragments causing discontinuity of the bone. A long tubular bone such as the humerus or femur usually fragments. These weight-bearing tubular bones require a higher degree of fragment reduction than do flat bones. In *incomplete fractures*, which occur commonly in immature bones, a portion of the cortex remains intact. Flat bones such as the scapula often "bend" instead of undergoing complete fracture, and may not undergo great fragmentation. Several other forms of fracture occur in mature bone as well. These may be described in several ways, including (1) greenstick, (2) buckling, (3) impacted, (4) fissure, or (5) penetrating. This classification can provide a technique for gauging the energy level of the injury that created the fracture. A fracture resulting from a low-energy injury might be incomplete, or complete with good apposition and alignment of the fragments and little soft tissue injury, and would be expected to heal readily, often with only minimal stabilization. A fracture resulting from a high-energy injury might be complete with comminution, fragment displacement, and more extensive soft tissue injury. These fractures heal more slowly and there is a definite need for reduction of the fragments and solid fixation. In very high-energy fractures, usu-

TABLE 3-1. Fracture Classification Based on Radiographic Examination

1. Nature of fracture
   a. Traumatic—single, violent force
   b. Stress or march—repeated stress
   c. Pathologic—weakened by disease (e.g., inflammatory, neoplastic, or metabolic disease)
2. Energy level of the trauma
   a. Low energy
   b. High energy
   c. Very high energy
3. Completeness of fracture
   a. Complete—fracture line extends through the bone involving the entire cortex (both cortical shadows are fractured as seen on a single radiographic view)
   b. Incomplete—fracture line does not involve the entire cross section of the bone and a portion of the cortex remains intact (only one cortical shadow is fractured as seen on one of the orthogonal radiographic views)
      1) Greenstick fracture—common in immature bones
      2) Buckling fracture—in bones weakened by a pathologic process
      3) Impacted fracture—in cancellous bone of the metaphyses as trabeculae are driven into each other; a distinct radiolucent fracture line is not identified
      4) Incomplete fracture—penetration of a foreign body (bullet)
      5) Stress fracture—repeated cycling
      6) Intraperiosteal fracture—cortical break with intact periosteum
4. Number of fracture lines
   a. Simple fracture—one fracture line
   b. Multiple fracture—more than one fracture line in the same bone organ, in which the fracture lines are not continuous
   c. Comminuted fracture—complete fracture with multiple small bone fragments in addition to the major fragments
   d. Segmental fracture—complete fracture characterized by a diaphyseal fragment with a complete cortex; the segmental fragment may be split longitudinally
5. Direction of the fracture line
   a. Oblique
   b. Spiral
   c. Transverse
   d. Fissure or longitudinal
   e. Saucer
   f. Avulsion
   g. Depressed
6. Location of the fracture line
   a. Diaphyseal
   b. Metaphyseal
   c. Epiphyseal–metaphyseal
   d. Epiphyseal (or articular) involving subchondral bone
   e. Physeal (skeletally immature patient)
7. Character of fracture fragments
   a. Impacted—compressed
   b. Comminuted—multiple fragments
   c. Butterfly fragment—large cortical fragment at fracture site
   d. Depressed—concavity in flat bone
8. Relationship of the fracture fragments
   a. Degree of end-to-end apposition of fragments
   b. Alignment of the fragments relative to adjacent joints and weight-bearing surfaces
   c. Degree of angulation between the fragments

(Table continued on following page)

TABLE 3-1. (*Continued*)

    d. Degree of rotation between the fragments
    e. Presence and extent of overall bone or limb shortening
  9. Soft tissue injury
    a. Closed fracture with no break in adjacent skin
    b. Open fracture with a break in adjacent skin characterized by
      1) Free air within soft tissues
      2) Bone fragments protruding through a break in the skin
      3) Radiopaque debris within soft tissues due to
        a) Surface contaminants
        b) Foreign body—bullet
    c. Interposition of soft tissues between bony fragments
    d. Swelling due to edema–hematoma
10. Associated joint injury
    a. Fracture line entering joint space
    b. Combination of fracture–luxation
    c. Avulsion fractures
    d. Articular corner fractures
11. Character of the reduced fracture
    a. Stable—fragments interlock to achieve stability
    b. Unstable—fracture fragments are such that they can rotate or slip on each other

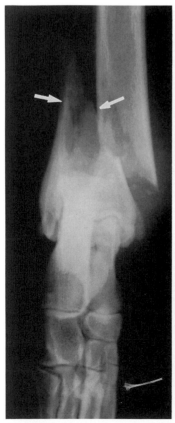
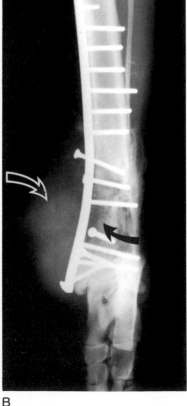

A   B

FIGURE 3-1. Craniocaudal radiographs of a fracture in the distal tibia in a 6-year-old male Irish Wolfhound. The first radiograph was made at the time of injury (A) and the posttreatment radiograph was made 60 days later (B). Cortical destruction (arrows) indicating underlying bone disease was not appreciated on the first study, and the fracture was treated using interfragmentary screws and a bone plate. The bone destruction associated with the osteosarcoma progressed (solid arrow) and amorphous tumor bone was noted within the soft tissue lesion laterally (hollow arrow).

ally caused by gunshot wounds, there is both soft tissue and bone loss. Treatment of the soft tissue injury is important, and the chances of infection and nonunion are high.

Injuries to all bones can be classified under two major subdivisions, intra-articular and extra-articular. *Intra-articular fractures* involve the articular surface and assume greater clinical importance, with anatomic reduction a definite goal as well as removal of all fracture debris from within the joint capsule. These fractures may be classified as: (1) linear with a single fracture line through the epiphysis into the joint, (2) comminuted with injury to a portion of the epiphysis, (3) impacted with collapse of a portion of the articular surface, and (4) having bone loss in which there is avulsion of a portion of the articular bone (Fig. 3-2). Articular fractures obviously may involve the proximal or distal end of the bone. If the fracture involves an articular surface, the technique of fragment reduction and the success in maintaining the reduction during healing assume considerable clinical significance. It is possible to combine an articular fracture with dislocation or an intracapsular fracture with a dislocation, creating a traumatic injury with a more severe prognosis.

The *extra-articular fractures* are conveniently divided into: (1) those involving the metaphyseal portion, and (2) those involving the shaft or diaphyseal portion. The number of fractures, direction of fracture lines, and degree of fragmentation create the possibility for a large number of fracture types. Assuming apposition, alignment, and rotation of the fragments in a nonarticular fracture are within or near anatomic limits at healing, the actual appearance of the healed fracture is of little clinical importance. This differs markedly from articular fractures, in which anatomic reduction of the fragments is most important. It is possible to have a third type of fracture: An *intracapsular fracture* is not articular but is important because of the possibility of loss of blood supply to the intracapsular fragment.

Shaft fractures may be located as occurring within the proximal one third, middle one third, or distal one third of the bone, and include four categories: (1) linear, (2) comminuted, (3) segmental, and (4) those with bone loss (Fig. 3-3). The fracture line(s) in linear fractures may be described by the *direction of the fracture lines*, and include: (1) transverse, (2) oblique, and (3) spiral (Fig. 3-4).

The *number of fragments* varies; fragmented fractures are described as those that are: (1) comminuted, (2) with a butterfly fragment, or (3) segmental. *Comminuted fractures* are divided into

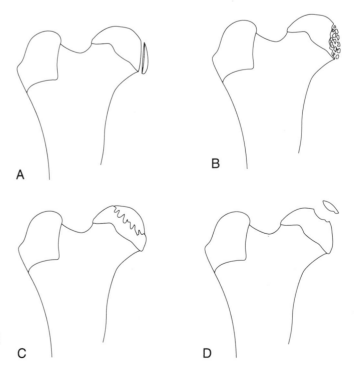

FIGURE 3-2. Drawing of the femoral head showing types of articular fractures in which the fracture is linear (A), comminuted (B), impacted (C), or with an avulsed fragment (D).

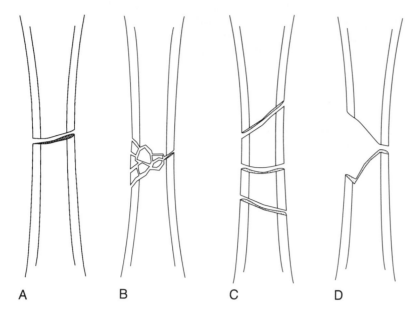

FIGURE 3-3. Drawings of the midshaft of a bone that illustrate categories of fractures: linear (A), comminuted (B), segmental (C), and those with bone loss (D).

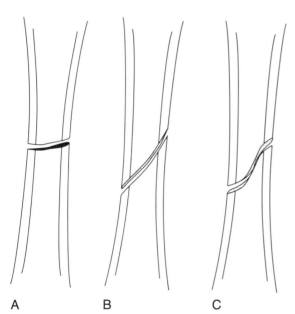

FIGURE 3-4. Drawings of the midshaft of a bone that illustrate categories of fractures as they are characterized by describing the direction of the fracture line: transverse (A), oblique (B), or spiral (C).

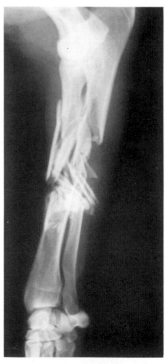

FIGURE 3-5. Lateral radiograph of the antebrachium of a 4-year-old male German Shepherd Dog with severely comminuted fractures of the radius and ulna in which the fragments originate from the diaphyses of both bones.

those in which the multiple comminuted fracture fragments originate in less than 50% of the shaft diameter or in 50% or more of the shaft diameter (Fig. 3-5). The nature of the comminution may be characterized by a single, larger *butterfly fragment.* Butterfly fragments are frequent in fractures of the dog and cat and consist of rather large cortical fragments that are smaller than the diameter of the bone but often large enough to permit placement of a lag screw. A single butterfly fragment may be smaller than 50% of the shaft diameter or larger than 50% or more of the shaft diameter. *Segmental fractures* consist of bone injury in which two (or more) fractures of the shaft are present, with a section of tubular bone creating a separate tubular fragment. It is also possible for the interposed tubular fragment(s) to be divided by a longitudinal split or become comminuted (Fig. 3-6). The extent to which the cortical diameter is involved by the separated fragment affects the strength of the bone after reduction and stabilization of the fragments. Identification of the categories of fracture greatly influences the character of the stabilization device selected.

Bone loss can be estimated and may be an important finding in the event of gunshot wound or a traction-type injury. Bone loss may be less than 50%, equal to or greater than 50%, or may be complete. The higher the percentage of bone loss, the more severe is the injury with regard to loss of bone tissue and the degree of soft tissue injury (see Fig. 3-3D).

Any of these fractures may be complicated by *fragment displacement,* so that there is overriding of the fragments. Most fractures in animals are complicated by the continued use of the limb after injury, so that fragments have no end-to-end apposition at the time of clinical examination. This secondary injury is a greater problem in patients with fractures of the pelvic limbs. In contrast, undisplaced fractures may occur in an immature animal after a minimal degree of trauma (low-energy fracture) with essentially 100% end-to-end apposition of the fragments. These fractures present the clinician with a much less severely injured patient, and there is a good opportunity for complete healing within 2 or 3 weeks. Remember that two radiographic views are necessary for determination of the nature of fragment displacement (Fig. 3-7).

A fracture may be *open or closed.* Diagnosis of an open fracture can be made by identification of a protruding bone fragment; however, it is more

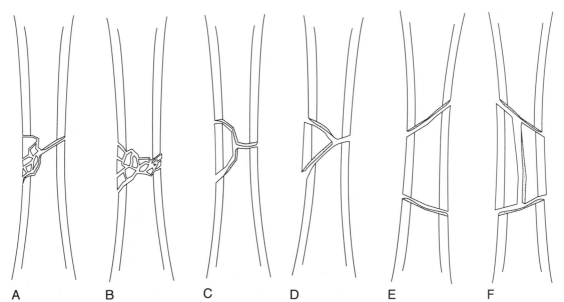

A   B   C   D   E   F

FIGURE 3-6. Drawings of a bone midshaft illustrating variations in the number of fracture fragments. Comminuted fractures are divided into those in which the fragments originate in less than 50% of the shaft diameter (A) or in 50% or more of the shaft diameter (B). Butterfly fragments may be smaller than 50% of shaft diameter (C) or larger than 50% or more of shaft diameter (D). Segmental fractures may have an intact interposed section of tubular bone (E), or the interposed fragment may be split longitudinally (F).

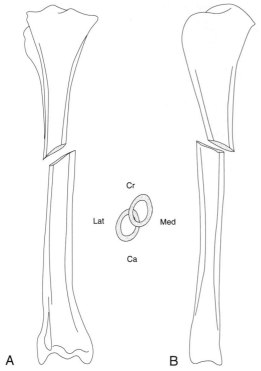

FIGURE 3-7. Drawings illustrating the need for two radiographic views to determine the apposition of fracture fragment ends. A craniocaudal view (A) suggests slightly more than 50% end-to-end apposition of the fragments, and the lateral view (B) suggests 50% end-to-end apposition. The cross-sectional drawing, however, shows the actual relationship of the fragments to involve only 25% of the cross section of the bone.

commonly diagnosed radiographically by identification of gas distributed in the subcutaneous tissues, within damaged muscle, or within fascial planes. The presence of radiopaque debris within soft tissues may be the result of a gunshot injury or may be road gravel; both suggest an open fracture. The possibility of infection developing within the soft tissues or bone is much greater with an open fracture. Debridement of injured soft tissues and flushing of the wound with copious amounts of fluid plus the initiation of antibiotic therapy are methods used to control the possibility of infection. Osteomyelitis is not always a sequela in open fractures, and is not always avoided in closed fractures.

Open fractures can be classified as first, second, or third class (or degree), depending on the amount of soft tissue injury. Grade 1 indicates that the bone fragment(s) penetrates the skin and the patient is presented within several hours. Because of the minimal soft tissue damage, this injury can be treated as a closed fracture. Grade 2 indicates that the soft tissue injury was from the outside. Contamination of the wound is possible and soft tissue injury is more severe than in Grade 1. Grade 3 makes reference to a wound that occurs from the outside with comminution; extensive skin, muscle, and possible nerve damage; and with a much more guarded prognosis. The risk of infection (osteomyelitis) or nonunion is much greater than in the other two grades.

Gunshot Injuries

Gunshot injuries may be common, depending on the culture of the society. The missiles include: (1) a small "B-B" or air-gun pellet, (2) numerous shot from a shotgun, or (3) a single high-velocity missile. Most of the lower-velocity missiles cause only soft tissue injury, although an air-gun pellet fired at close range can cause a fracture of a long bone in a cat (Fig. 3-8). Evaluation of the radiographs can contribute to an understanding of this type of injury (Messmer and Fierro, 1986).

Low-velocity missiles such as occur from shotgun wounds are seen commonly on the radiograph with most of the shot pellets remaining within the subcutis. However, if the injury follows a close-range shotgun blast, massive destruction of soft tissue and bone may occur.

The *high-velocity missile* from a hunting weapon causes extreme tissue injury, and resulting fractures usually are highly comminuted with associated severe soft tissue injury. The wounding effect of such missiles varies, depending on mass, shape, velocity, deformation, and whether the missile tumbles or breaks. Velocity is the most important factor. When a bullet strikes a solid object, all or part of its kinetic energy is transmitted to the tissue. Particles of bone accelerate forward and act as secondary missiles. One of the primary features of all missile wounds is cavitation. Within milliseconds after a high-velocity missile impacts and perforates, a pulsating, undulating, temporary cavity is formed. The surrounding tissue is subsequently explosively pushed and compressed laterally to enclose the temporarily formed cavity. The maximum diameter of this temporary cavity may be approximately 30 times the size of the original missile track. Therefore, tissues at a distance

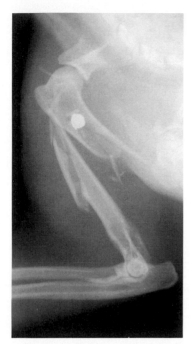

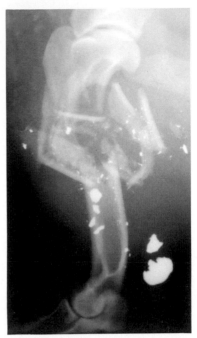

FIGURE 3-8. Lateral radiograph of the humerus in a 3-year-old male, castrated, domestic longhaired cat with a comminuted fracture resulting from an air-gun pellet.

FIGURE 3-9. Radiographs of a 6-year-old male Golden Retriever with a comminuted humeral fracture resulting from a high-velocity missile that fragmented on contact.

from the original wound may be damaged and adjacent bones may be fractured without ever having been struck directly by the missile. In contrast, lower-velocity missiles create a direct pathway of destruction, with little injury to surrounding tissues.

If the missile is an expanding type (hollow tip), it mushrooms on impact with tissue, so there are multiple metallic fragments. Thus, the *bullet track* can be identified from the pattern of deposition of variously sized lead fragments made as the bullet passed through the soft tissues (Fig. 3-9). If the bullet is steel coated or has a hard coating of another type, there may be no fragmentation. The bullet track through the patient's body should be determined either radiographically or clinically, so that other organs that are suspected of injury can be identified and evaluated in addition to the fractures seen radiographically.

The *healing of fractures* due to gunshot wounds depends primarily on the ingrowth of a new blood supply from the damaged soft tissues. Thus, the

bony and metallic fragments, even when left in position, do not seem to affect healing adversely to the degree that might be expected.

Physeal Fractures

Physeal fractures in the skeletally immature patient can be classified as Salter-Harris types I, II, III, IV, and V, and a separate Rang type VI, a grinding type of injury with bone tissue missing from both metaphysis and epiphysis (Salter and Harris, 1963; Fig. 3-10). These fractures are unique injuries seen frequently in the skeletally immature patient, in which the growth region of the long bone is damaged and there is the potential for alteration of the bone's normal growth pattern. The classification of physeal fractures is well described on the basis of involvement of the physis, metaphysis, and epiphysis (Rogers, 1970; Kleine, 1971; O'Brien, 1971; O'Brien et al, 1971). The radiographic diagnosis of physeal injury is less well described (Kleine, 1971; Morgan, 1972).

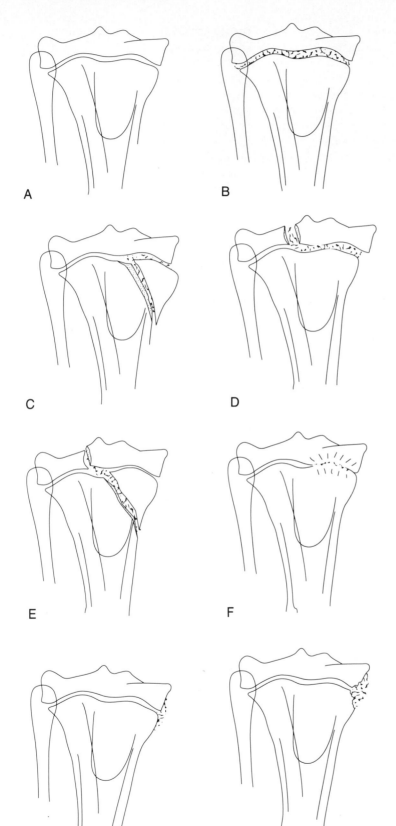

A

B

C

D

E

F

G

H

FIGURE 3-10. Drawings of the proximal tibia as seen on a craniocaudal radiograph, illustrating physeal injury: normal physis (A), Salter-Harris type I (B), Salter-Harris type II (C), Salter-Harris type III (D), Salter-Harris type IV (E), Salter-Harris type V (F), and a Rang type VI, in which a soft tissue injury bridges the physis (G) or in which there has been an abrasive injury affecting the epiphysis and metaphysis (H).

In the *simplest form (type I)*, a physeal fracture consists of injury to the replicating cartilaginous cells in the physis, with the possibility of the bony fragments returning to their normal anatomic positions after the injury or remaining with various degrees of malposition. This type of injury occurs within the distal radius, proximal femur, and the proximal humerus, in which the physis is plate-like and separation can take place without injury to the bony metaphysis. It repairs relatively well after anatomic reduction because there has been little destruction to either the metaphyseal or epiphyseal blood supply and no opportunity is present for cross healing between fragments. Radiographic evaluation of the healing is difficult, because no bridging callus is present. If a bridging callus does form, it would prevent further bone growth.

A *type II physeal fracture* is a more severe injury to the bone (physis), and results in displacement of the fragments. The fracture line escapes from the physis, runs into the metaphysis, and separates a small, triangular bone fragment from the metaphysis. This injury occurs in physes that have an undulating surface, often with four pyramid-shaped protrusions that fit into four depressions, giving strength to the growth region, and is found within the distal femur, distal humerus, and proximal tibia. Callus forms around the metaphyseal fragment and can be noted radiographically. The growth plate may close with this type of injury because of the possibility of cross healing between fragments.

A *type III physeal fracture* involves the physis in part, but the fracture line turns and passes through the epiphysis. This fracture is articular in addition to involving the growth plate. It is less common, but is seen in the distal humerus, the distal radius, and the proximal tibia. Anatomic reduction is a requirement if the patient is to have complete use of the joint. The fracture does not injure the blood supply to the metaphysis, and growth plate injury is uncommon.

A *type IV physeal fracture* involves the physis in part. The fracture line is more longitudinally directed and passes through the epiphysis, across the physis, and into the metaphysis, exiting a short distance from the physeal plate. This fracture is common in the distal humerus and distal femur, and is cause for great care to be taken in its treatment. If the fragments are not returned to anatomic position, it is possible for a bony bridge to form between the metaphysis and epiphysis, affecting further bone growth as well as destroying the articular surface.

A *type V physeal injury* occurs when the trauma causes a crushing injury to the growth plate and periphyseal soft tissue. Bony bridging may occur, preventing normal longitudinal bone growth. This fracture occurs commonly in the distal ulna, where the shape of the epiphysis prevents the metaphysis from escaping during trauma with resulting crushing of the physeal plate. Delayed growth or premature closure of the growth plate results, causing clinical problems because of the normal growth of the paired radius.

A *Rang type VI physeal injury* occurs when there is soft tissue trauma adjacent to the physis that generates a fibrous or bony scar that can bridge the physis, influencing future bone growth. This type of injury can be made more severe by being an abrasive one that destroys bone tissue within the epiphysis and metaphysis adjacent to the physis. This injury often occurs distally in the limbs and involves the distal radius–ulna or the distal tibia–fibula. The scar may bridge the growth plate and severely alter bone growth (Rang, 1975).

An important consequence of any type of bony injury involving the physes results from the possible *interruption of normal bone growth*. The interruption may cause: (1) premature closure within a growth plate with resulting cessation of growth; (2) an interruption of growth that may be temporary, resulting in a shortened bone, with normal or near-normal growth later; or (3) an interruption or cessation of growth that involves only a part of the affected physis, resulting in angular deformity. These injuries to the growth potential of the bone may result in shortening of the affected bone with or without angular deformity, depending on the part of the physis affected and the duration and extent of the effect. Obviously, a paired bone adjacent to the affected bone may be affected by the injury, and the magnitude of deformity is greatly increased (Clayton-Jones and Vaughan, 1970; Newton, 1974; Ramadan and Vaughan, 1978; Marretta and Schrader, 1983; Fox, 1984).

Apophyseal Fractures

Avulsion of apophyseal growth centers is a rather frequently occurring injury after trauma in which the

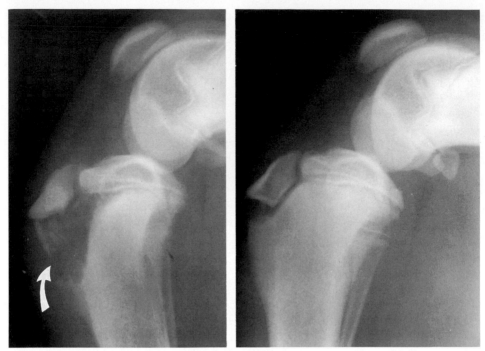

FIGURE 3-11. Lateral radiographs of a skeletally immature dog with an avulsion fracture of an apophyseal growth center (arrow). The radiograph of the normal limb (right) is included.

bony fragment is displaced proximally away from the bone. Injury of this type commonly is associated with the supraglenoid tuberosity, greater trochanter of the femur, the tibial crest (Fig. 3-11), and the growth center for the olecranon process (Pond, 1975). Additional traction-type injuries involve the growth centers of the accessory carpal bone, the calcaneus, the lesser trochanter of the femur, the humeral tubercles, and the medial epicondyle of the humerus. No predisposing reason for the separation exists and the injury is not an indication of underlying bone disease. Because it is a trauma to healthy bone, the injuries usually are unilateral. Usually the bony portion of the apophyseal growth center does not fracture because the cartilage plate separating the apophysis from the parent bone is the area of greatest weakness, and the fracture lines are limited to that region. Because the apophyses in the immature patient usually do not fragment with avulsion, they can be reattached to the parent bone through use of small metallic screws or K-E wire(s) and stabilized through use of a tension-band device during healing.

Because apophyses do not contribute prominently to the length of the bone, are not articular,

and do not require a specific shape to function as a tendinous attachment, perfect repositioning of the avulsed fragment that leads to renewed physeal growth is not required. The possibility of joint injury is remote because this type of injury is away from the articular surface and extracapsular in location. Every effort should be made to keep the surgical repositioning of the apophyses extracapsular as well. Because of the available soft tissue blood supply, most apophyseal center avulsion fractures heal easily after repositioning. Separation of the tibial crest is a unique injury and is important clinically because it alters the length of the patellar ligament and may influence the femoropatellar joint. Repositioning of the avulsed fragment should be made with a view to reducing the patellar luxation so that it articulates normally with the trochlear groove of the distal femur.

Repeated clinical and radiographic examinations are necessary in the evaluation of potential growth abnormalities because of the rapid rate of growth in the long bones. The time between the original trauma causing the physeal injury and signs of delayed growth or cessation of growth depends on the age of the patient, the rate of bone growth, and the nature of the injury. Often the most obvious radiographic changes indicating a problem of bone growth are seen within a paired bone between 2 and 4 weeks posttrauma. The time required for change to be noted within a single bone is less well studied. Thus, unless the owner is advised of this potential growth problem, injury to the adjacent joints because of bone angulation or bowing may become advanced before the owner seeks medical assistance. Correct diagnosis of the original traumatic injury or detection of early secondary changes obviously is important in determining the prognosis for a patient only beginning to show abnormalities, and for determining the timing and need for orthopedic surgery in the more severely affected patient.

Remember that identification of the fracture is only the first part of the radiographic diagnosis. A complete description of the fracture type permits a determination of the type of stabilization required and a prognosis for healing to be made. During healing, the fracture may be classified as *stable or unstable*. A stable fracture is one in which the fragments interlock and provide resistance to collapse and some resistance to rotation. An unstable fracture is one in which the fragments do not interlock and thus are free to slide when loaded.

Periosteal Injury

It is important to estimate *periosteal injury*, because fracture healing occurs primarily through extracortical new blood supply (the extraosseous blood supply of healing bone). If the periosteum is stripped, the blood supply to the underlying bone is compromised because of the development of a large subperiosteal blood clot. This delays external callus formation because it separates the new blood supply from the underlying bone. Periosteal stripping occurs to a greater degree in the immature patient because of the loose periosteal attachment to the underlying bone. This tearing creates a bizarre pattern of callus formation around the fracture.

Soft Tissue Injury

Distinction between *strain, sprain, and dislocation* is difficult to make radiographically and is assisted principally through the use of stress radiography, which demonstrates the extent of soft tissue injury and the resulting level of joint instability.

Orthopedic Fixation Devices

4

Introduction

Orthopedic fixation devices are used in the treatment of fractures, soft tissue injuries, and reconstructive surgery, and have been described earlier (Lawson, 1963). Their radiographic appearance is not described often (Morgan, 1972, 1981; Manley et al, 1981; Richardson et al, 1987; Slone et al, 1991). After fracture reduction, internal, external, or intramedullary fixation devices may be used to reduce the fracture and provide stability and maintain the alignment of bone fragments during the healing process. Coaptation splints and casting are the oldest methods used for fracture treatment, but are not often recommended today because of advances in treatment techniques. Screws are used primarily to provide interfragmental compression or to attach plates, which can then provide compression in addition to preventing displacement and supporting the fragments during healing. Pins and wires can be used for fixation of small avulsion fragments, reduction of small bone fragments, for attachment of external fixation devices, and for intramedullary placement in long bone fractures.

The use of these devices became possible only with the discovery of x-rays, availability of anesthetics, and an understanding of surgical asepsis. Selection of the technique of reduction and fixation depends on the patient, type and location of the fracture, associated injuries, and experience of the surgeon. It should be remembered that an excellent result can be obtained in treatment of fractures without "anatomic" reduction of each fracture fragment, as long as the overall alignment of the bone is restored. Anatomic apposition is preferred, but is not always necessary for good clinical healing.

The *placement of the orthopedic devices* should be evaluated carefully on the postoperative radiographic study. Although the intention behind radiography is not to point out every slight error on the surgeon's part, it is important to look for errors that lessen the strength of the stabilization. A decision that the reduction is not satisfactory needs to be made immediately so that the use of additional devices can be considered. The decision to repeat a reduction completely needs to be considered while the fracture is still fresh instead of waiting several weeks, by which time an easy surgical exploration of the fracture site with relatively simple identification of the fragments is impossible.

Reduction is divided into *closed reduction*, in which the functional alignment of the bone can be restored and fixation achieved without exposure of the bone fragments, or *open reduction*, for treatment of fractures that require reduction that cannot be achieved with closed techniques. Fixation devices also are described according to whether *compression* can be produced across the fragment surfaces. With adequate compression, movement is eliminated and the resulting callus formation is small, whereas with other fracture devices stabilization is achieved without compression and the bridging callus is larger.

Coaptation Splinting and Casting

The oldest, simplest, and least expensive means of fracture treatment are *splints and casts*. They may be used on incomplete fractures, such as a greenstick fracture of the radius in a puppy, as a temporary control to prevent further injury during transportation or emergency treatment, or as a protection for internal fixation. They often are ineffective and actually can be harmful in the treatment of complete long bone fractures. Pressure sores, toe necrosis, joint stiffness, muscle atrophy, osteopenia, prolonged healing time, nonunion, malunion, the necessity of close observation, and frequent repair and replacement often negate any benefit of their low initial cost.

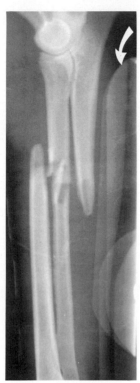

FIGURE 4-1. Lateral radiograph of the antebrachium of a dog with a fractured radius and ulna. Note the dense shadow cast by the Mason metasplint (arrow).

Coaptation splints are made of board, aluminum (Mason metasplint; Fig. 4-1), and plastic. Board splints usually are applied to fractures of the radius and ulna. A full-length board splint may be used to immobilize the entire forelimb. In the pelvic limb they are applied to stabilize fractures of the tibia, tarsus, and metatarsal bones. Aluminum splints and plastic splints are used in the forelimb in the management of fractures of the radius and ulna, carpal bones, and metacarpal bones.

A special traction splint, the *Schroeder-Thomas splint*, consists of an aluminum rod shaped to form a ring that fits into the axilla or groin with two extensions distally that are bent and end in a "U"-shaped piece at the foot. The rods that extend from the ring are angled to reflect normal flexion of the joints of the limb. Bandages and adhesive tape are applied to bind the limb to the splint and immobilize it. The splint may be used as a temporary protection during transportation and

emergency life-saving procedures, but it is not recommended as a treatment of choice for long bone fractures. It does not provide enough stability, requires constant revision and care, and can be harmful if not correctly applied and maintained. Remember that the splint places the limb in traction, whereas other forms of fracture repair create compression at the fracture site.

The *Robert Jones splint* is highly recommended as emergency protection for a fractured limb, as an adjunct to internal fixation after the surgery, and after plate removal. It provides the most protection with the least chance of doing harm. The application is simple and it is well accepted by the patient. Adhesive tape is applied to the lower limb and is used for traction while thick layers of cotton are wrapped in a spiralling manner distal to proximal. Four inch conforming bandage is then wrapped over the cotton padding. The first layer of bandage provides slight compression of the cotton; the second layer further compresses the cotton padding. Foot tapes are then reflected onto the splint to prevent it slipping distally. A final covering of Vetrap® is applied very firmly.

Casts of plaster of Paris have been used in treatment of fractures for over 100 years. However, modern-day plastic casting materials such as fiberglass are much superior. Casts are not very satisfactory as the primary method of treatment, although they can be useful as a temporary adjunct to distal long bone fractures treated with internal fixation. They are especially useful in providing support in carpal and tarsal fusions. They require considerable care in their application and observation to prevent: (1) pressure sores, (2) loosening, (3) becoming wet and soft, and (4) becoming soiled.

With the advent of superior methods of fracture repair, splints and casts, which were once the sole means of treatment, should be used with caution and only where they are beneficial. Today, this usually is limited to emergency situations and in an auxillary fashion to provide support for another method of fixation.

Radiographic interpretation of techniques chosen for fracture repair or of the status of fracture healing is made difficult by the presence of most casts and splints, because they create radiopaque shadows that often are superimposed over the fractured bones, and often only one unobstructed view is available for examination (see Fig. 4-1). Casts and metallic splints make the visualization of early healing callus especially difficult, and the radiograph should be used only to determine apposition and alignment of the fragments.

Screws

Cortical and *cancellous* are the two basic types of screws available. Cortical screws are threaded over their entire length, usually have blunt ends with shallow threads that are closely spaced, and are designed to achieve maximum purchase in dense cortical bone. Cortical screws can be used as lag screws (Fig. 4-2). Cancellous screws have a wider thread diameter with a steeper pitch, a large area between threads to improve the screw's purchase in cancellous bone, and are used principally in metaphyseal and epiphyseal fractures. They can be partially or completely threaded, and also may serve as lag screws. Special screws are hollow and are inserted over small-diameter guide pins, which are removed after the screws are placed. Both cortical and cancellous screws are available as cannulated screws.

Screws convert torque into compression and are used primarily to provide interfragmental compression across a fracture site, as well as to attach plates to bone. A screw that crosses a fracture line should be placed as a lag screw, meaning that only the threads on the "far" fragment gain purchase. Only in this way can the far fragment be drawn tightly against the near fragment. The surgeon can select a fully threaded screw and overdrill the near fragment (create a glide hole), or use a partially threaded screw to obtain compression. Both cortical and cancellous screws can be used for compression, with the selection depending on the nature of the bone tissue. If the screw is used in cancellous bone, a washer can be used beneath the screw head to distribute the force of the head more widely on the softer bone. These screws do not provide sufficient fixation to protect most fractures from the normal bending, rotation, and axial-loading forces encountered by the bone during healing, and these forces must be neutralized, usually with a plate (used in a neutralization mode). Multiple screws may be used to provide greater rotational stability. It is possible to use a screw only as a positional screw in that it simply holds a fragment in position while an additional stabilization method is applied.

Radiographic examination of postoperative frac-

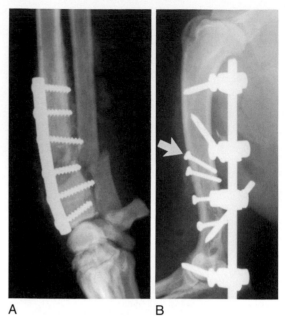

A   B

FIGURE 4-2. Lateral radiograph of the antebrachium of a dog with fractures of the radius and ulna, with a bone plate held by cortical screws stabilizing the radial fracture (A). Lateral radiograph of a humeral fracture in a dog (B). The fracture was stabilized by cortical screws used in a lag fashion (arrow) and by a type I external fixator.

tures should evaluate screw placement, especially the number of cortices that are engaged. The position of the fragments should be evaluated because the more tight the placement, the earlier the fracture heals. However, if a glide hole has not been properly drilled, screw placement actually can displace fracture fragments and hold them at a distance from other fragments, and this separation needs to be identified and a decision made relative to correction.

Screw placement is of particular importance and needs special attention. The proximal screws within the femur often should be directed into the middle of the femoral neck, and their length should be evaluated. If the screw is longer, note the relationship of the screw tip relative to the articular surface of the femoral head, and ensure that it has not entered the joint. An important use of transcondylar screws is within the distal humerus, and their placement needs to be judged relative to location and fragment compression. The available space for a screw used in this manner in the cat is much less than that in the dog, and avoiding the articular surface is critical. The use of long screws within the radius should be carefully considered because they may incorporate the ulna as well and lead to synostosis between the two bones (see Fig. 4-2). This actually may be desired in certain fracture treatments. Screws can be used for interfragmentary compression (Nunamaker, 1973). Placement of interfragmentary screws is difficult to evaluate on the radiograph because of the selection of varying angles of placement by the surgeon. The screws should be positioned at an angle created by dividing the arc formed between a line perpendicular to the long shaft of the bone and a line perpendicular to the fracture line. The importance of this placement may have been overstressed. Regardless of its importance, determination of the angle usually is difficult radiographically.

Screw placement with a bone plate needs to be evaluated carefully to determine how many cortices have been penetrated proximal and distal to the fracture site (see Fig. 4-2). In the event of long oblique fractures, it may be difficult to penetrate a sufficient number of cortices within the proximal and distal fragments, and the fracture may be left unstable after plate positioning. Screws sometimes are seen where the far cortex has been overdrilled or the far threads have been stripped. In this situation, a nut with or without a washer can be placed on the end of the screw to strengthen its hold on the bone. To avoid creating a weakened "line" in the bone, the screw holes are drilled at varying angles. The different angulations of the screws are easily recognized on the radiograph that positions the plate "en face." However, on the radiograph that visualizes the plate "on end," screws that are placed within different planes appear to be of varying lengths, and some actually may appear to not penetrate the far cortex. It is best to study the longest screw and, noting that it penetrates the far cortex, assume that adjacent screws are of the same length and actually penetrate the far cortex as well. Their shorter appearance is probably due to their placement at an angle relative to the x-ray beam. This is a situation in which one can rely on the skill of the surgeon and assume that all is well.

Bone screws that are small for plate holes appear to be loose relative to the plate. Screws with large heads relative to the plate holes seem not to have as tight a fit as required. These situations are difficult to evaluate radiographically.

Screw placement near a growth plate must be considered carefully. Screws should not cross an open growth plate because the effect is that of preventing growth by fixing the epiphysis and the metaphysis under compression and preventing their separation secondary to normal growth. Because the growth of an apophyseal plate is not as critical, a screw can be used that crosses this growth plate in the repositioning of a traumatically avulsed or surgically osteotomized apophyseal center.

Screw appearance in a healing fracture should be evaluated on later radiographic studies to identify screws that are loosening, those that have pulled out of the bone, or those that have broken when subjected to excess loads. Screws in an infected environment show increased lucency in the bone around the screw, with localized pockets of osteolysis. Uncommonly, there may be a generalized lucent zone around a screw that is due to movement of the screw without infection. Protruding screws that have backed out can be recognized radiographically and may cause irritation of the soft tissues without any other radiographic changes except for soft tissue swelling.

Bone necrosis can occur in association with production of excessive heat at the time of the original drilling of the screw holes. If this occurs, the bone tissue adjacent to the screw retains its density because it has become avascular, whereas bone lysis occurs within the viable bone tissue just adjacent to the dead bone. This creates a separated rim of dense bone around the screw that can be seen radiographically. This pattern of change is commonly seen in more dense bone tissue such as is found in the horse, but is a relatively uncommon finding in the dog and cat, and then only in larger breeds of dogs with more dense bone.

Plates

Plates come in various sizes and shapes and are used in different functions to compress, neutralize, or buttress a fracture (Fig. 4-3). *Compression plates* are used in reduction and stabilization of fractures that are stable when placed in compression; however, the plates may be used in combination with lag screws if separate bone fragments are present. Certain bones have a compression or tension side, and a plate placed on the tension side absorbs the tensile forces, resulting in dynamic compression of the fracture. The *limited contact dynamic compression plate* has been developed to minimize vascular damage to the plated bone segment. Plate contact

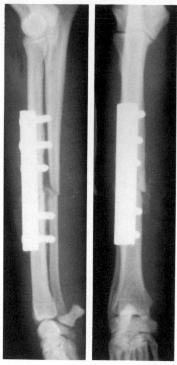

FIGURE 4-3. Lateral and craniocaudal radiographs of a dog with fractures of the radius and ulna. Stabilization of the radial fracture was accomplished through a six-hole plate used in a compression mode.

the stress through the bone. Plates also may be used to buttress a damaged articular surface. Cancellous grafts often are used with plates used to buttress because it is necessary to fill the gap created by bone loss as quickly as possible.

The plates may be straight with *round holes, or may have unique oval holes* with inclined edges that permit them to function as dynamic compression plates. The screw head may be placed so as to glide down the inclined edge, moving the plate relative to the bone. The screw also may be placed in a neutral position within the screw hole with no intent to shift the position of the plate relative to the bone.

Plate contour needs be noted carefully on the radiograph because the position of the plate relative to the surface of the bone provides information as to the possibility of the plate cycling due to motion, with a resulting fatigue fracture. If the radiographic positioning is such that the plate is seen "on end," it is relatively easy to note how tightly the plate fits the bone. A poorly contoured plate has a radiographically lucent zone between the plate and the bone. However, both radiographic projections often are made at an angle relative to the plate so that it is not possible to visualize the fit of plate onto the bone, and identify the space between the bone and the plate.

Plate length and weight are both important, but this selection should have been carefully evaluated before surgery. It is to be hoped that the radiograph will not bring to light the fact that the plate is of improper length or size.

Tubular plates or semitubular plates are thin and have a concave inner surface that conforms to the curvature of the bone surface; they are pliable and easier to contour. *Contoured plates or reconstruction plates* are designed to allow bending, twisting, and contouring to accommodate bones with unusual shapes such as the acetabulum. Other special plates have a unique shape such as *T or L shape*. Small plates are called *miniplates or finger plates*.

Plating alters the manner of fracture healing and consequently the way the healing fracture appears radiographically. With minimal movement at the fracture site, callus formation is rather exuberant, and its presence is a good radiographic indicator of fracture healing that can be evaluated both as to the time of its appearance and the amount of callus formed. With anatomic reduction and rigid fixation (compression), the amount of callus for-

with the underlying bone surface has been reduced as much as is possible. This plate provides the same rigid stability as a regular plate, yet avoids, as much as possible, disturbances of the bone in the vicinity of the implant. This helps to improve the blood supply to the bone, avoids sequestra formation in the presence of infection, and confines possible infection. *Neutralization plates* only protect fracture surfaces from normal bending, rotation, and axial-loading forces, and often are used in combination with lag screws, which are usually, but not always, placed separately from the plate. The plate absorbs a significant portion of the stresses generated by weight bearing and thereby protects the reconstructed area of the bone from excessive loading and failure. Some plates both compress the fracture and neutralize the loading forces. *Buttress plates* support a bone that is unstable in compression or axial loading, and are used for repairing fractures with severely comminuted fragments in which there is large bone loss. Because of the loss of bone, the plate must bear all of mation is limited, with the possibility of a total absence of callus development as seen radiographically. Because this callus formation is expected, its absence may be of concern, and raises the suspicion of delayed healing or nonunion. In a patient with compression at the fracture site, the disappearance of the fracture line on the radiograph is a more important indicator of healing.

Fatigue fracture of the plate, one of the most serious complications, results from excessive motion at the fracture site with repetitive stress to the plate, and occurs in unstable implants subjected to prolonged cyclic loading. There is often a race between bone healing and fatigue failure of the plate. Although the fractured plate is easy to note radiographically, it is important to evaluate the technique of original fixation because it may be possible to predict insufficient stability at the fracture site and correct it to avoid cycling and subsequent plate fracture.

Stress protection is another problem that may result because the plate is much more rigid than the bone, and much of the stress of weight bearing is absorbed by the plate. In this situation, the cortex beneath and around the plate becomes osteopenic because there is no requirement for bone tissue to support the limb. Often the bone responds to the lines of stress in such a manner that they can be followed from the bone through the screws onto the plate and then back to the bone. This pattern will have a sclerotic appearance and will surround or contrast the areas of osteopenia in which there is no stress (Fig. 4-4). The bone atrophy can be noted radiographically as a thinning of the cortical shadow as well as a loss of bone density, and it may lead to a weakened bone and a new fracture of the bone after removal of the plate, after the original fracture has healed. The fractured bone usually requires some type of external support after removal of stabilization devices until it regains original strength.

Osteomyelitis in the bone beneath the plate leads to focal lucencies that are recognized only on a perfect lateral projection of the plate. The infection also may lead to loosening of the screws, permitting the plate to shift away from the bone. Bone infection associated with fracture healing usually is suspected clinically before radiographic evidence of infection. Evidence of infection is a reason for surgical removal of the screws and plate. Drainage tracks often are present in association

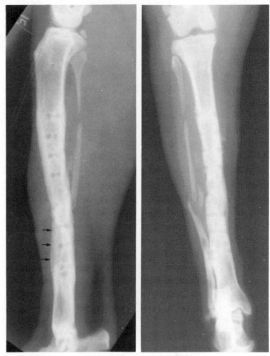

FIGURE 4-4. Radiographs of the tibia of a 3-year-old male, castrated, domestic short-haired cat with a tibial fracture treated by application of a bone plate. After removal of the plate, stress protection is evidenced by failure of cortical bone redevelopment. The lines of force passed from the bone through the screws to the bone plate, creating sclerotic patches around the screws. Note the lucency at the fracture site, where the most prominent stress protection is noted (arrows). A form of bone atrophy has occurred in the fibula, with failure of healing between fragments, because of the support provided by the plating.

with infected plating. Radiographs made after the injection of a liquid positive contrast agent into the sinus track may provide the surgeon with information as to which screw is the offender, or perhaps locate another cause for the soft tissue drainage.

The decision as to *plate removal* after fracture healing is based on several factors, and thus is made on an individual basis. Plate removal should be considered: (1) if stress protection is suspected, (2) if there is thermal conduction that is painful to the patient, (3) if mechanical irritation of overlying tendons or ligaments is suspected, (4) in the event of infection, or (5) usually in a young patient. Plate removal may be staged in the event of concern for the underlying strength of the bone. A

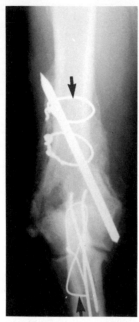

FIGURE 4-5. Craniocaudal radiograph of the distal humerus with two cerclage wires (arrow) used to reduce an oblique fracture before placement of a single intramedullary pin. Note the figure-eight wire (arrow) used in a tension-band device to reduce the olecranon osteotomy.

badly comminuted fracture may develop an exuberant callus as seen on the radiograph that "grows" over the plate, making its removal difficult. Usually, plates are left in position in the pelvis or scapula and in older patients.

Wires

K wires are unthreaded segments of extruded wire of variable thickness that are drilled into bone by placing the wire into a drill as if it were a drill bit. The wires provide rotational stability when more than one wire is used. K wires also are used for fixation of small bone fragments. The wires can be placed across a physeal plate without compromising the growth potential of the bone because the wires are smooth and the growing bone "slides" along the wire (Morshead, 1982). If two or more wires are placed so they cross a growth plate, they must be parallel so that the bone may slide over the wires. If the wires are placed in a crossing fashion, growth is inhibited. K wires also can serve as guide pins for the placement of cannulated screws. Wires also can be used in a transfixation

mode with a plaster bandage supporting the wires externally (Björck, 1952). Bioabsorbable polymeric composites have been used in the development of rods of varying diameter for internal fixation of cancellous bone fractures, osteotomies, or arthrodesis in small animals (Mero et al, 1989).

Cerclage wiring refers to the technique of placing monofilament stainless steel wire around the bone to stabilize fracture fragments, and usually is used in combination with other types of fixation devices (Hinko and Rhinelander, 1975; Withrow and Holmberg, 1977; Gambardella, 1980), but can be used alone (Fig. 4-5). Use of cerclage wires is indicated for reduction of long oblique fractures; they also can be used to reposition separate fragments, but the fragments must have anatomic reduction or they become unstable when the wires are tightened. In some fractures, butterfly fragments are reduced through the use of cerclage wires, and after anatomic reduction they are held more firmly with interfragmentary screws. Twenty-two-gauge wire is used for cats and small dogs, 20-gauge wire is used for medium-sized dogs, and 18-gauge wire is used for large and giant breeds of dogs. The radiographic examination must evaluate, if possible, whether the fragments are anatomically reduced and whether the wires are tightly positioned. The manner in which the two wire ends are twisted can be seen on the radiograph. If the wires are loose, they move along the shaft of the bone, disrupting the healing bone's new extraosseous blood supply that originates from the soft tissues, and a nonunion fracture frequently results (Newton and Hohn, 1974). The wires tend to move toward the fracture site, creating a fulcrum around which motion of the fragments tends to center. Because the width of the shaft of a long bone varies it is possible for the cerclage wire to move from the point of wider diameter, where the wire was tightly placed, to a point of smaller diameter, where the wire is loose and freely moves. This movement can be partially controlled by creating a shallow groove around the bone into which the wire fits, thus preventing it from shifting in location. This groove cannot be visualized radiographically beneath the cerclage wire. Placement of a small K wire transversely across the bone shaft can assist in holding the cerclage wire in position and prevent movement along the bone shaft.

Hemicerclage wires partially encircle the shaft of bone but then enter and exit through the cortex, passing through the medullary cavity. Use of the wire in this manner ensures that it cannot become loose and slip along the shaft of the bone. A hemicerclage technique often is used in a short oblique fracture in combination with an intramedullary (IM) pin, with the wire encircling the pin and holding it firmly against an endosteal surface.

The *location of cerclage and hemicerclage* wires needs to be evaluated carefully on radiographs made during healing of the fracture. Movement of the wires is a positive indication of instability at the fracture site, with a strong possibility of delayed union or nonunion. Cerclage wires often are seen to be broken on postsurgical radiographs made to determine the progression of fracture healing. The significance of the broken wires depends on the progress in the development of healing callus that stabilizes the fracture. If the fracture is determined to be healed by identification of a bridging callus, the presence of the broken wires is of little consequence. However, in the presence of only partial healing, the finding of broken wires is of great clinical significance, because it indicates the possibility of motion at the fracture site and the subsequent prevention of further healing. Callus cannot form beneath the cerclage wire because the wire prevents the entrance of the new blood supply (Göthman, 1962a–c). Occasionally, bone tissue on a radiograph of a healed fracture is noted to have formed in an encircling manner over the cerclage wires, making their removal almost impossible. The retention of the wire within the healing callus is not thought to be of clinical consequence.

Tension-band wiring is a unique technique used to provide dynamic compression for the treatment of avulsion-type fractures, such as those involving the olecranon, tibial crest, or patella, or for the replacement of apophyses that have been removed surgically to provide access to the bone or joint (Birchard and Bright, 1981). Parallel K wires are placed to provide rotational stability and reduce shearing forces between the fragments. The ends are cut and bent at right angles and turned into the tendon. A figure-eight wire is placed on the tension side of the bone and is anchored by passing it around both ends of the K wires and through a drilled hole in the bone (see Fig. 4-5). When physiologic forces pull on the bone, the wire carries the tensile force, which prevents separation, transmitting compressive forces to the bone. With time, the figure-eight wires may break, or the wires straighten, permitting them to slip free. This is a

FIGURE 4-6. Craniocaudal radiograph of the elbow that illustrates improper placement of tension-band K wires (arrows) used to stabilize the osteotomy. Because of the unique morphology of the proximal ulna, it is relatively easy to place the wire into adjacent soft tissues.

problem only when it occurs early in the course of fracture healing. Tension-band wiring rarely becomes associated with an infectious process.

K-wire placement associated with tension-band replacement of osteotomies is important to note radiographically, especially within the proximal ulna when tension-band wiring is used to stabilize the olecranon process. Because of the angulation of the proximal ulna in the dog, it is relatively easy to cause the wire to exit incorrectly from the bone. Unfortunately, it is also possible for the wire to enter the adjacent elbow joint space. These problems are especially important because the fracture may heal adequately, but the unnecessary injury to the joint may leave the patient without full use of the limb (Fig. 4-6). Correctly placed wires rarely are removed after fracture healing. If a wire is superficial and is causing a soft tissue lesion, there may be reason for its removal.

External Fixation Devices

External fixation is a technique in which individual bone fragments are held in place by transfixation wires or pins positioned percutaneously and attached to an external frame that may be metallic or plastic (Sisk, 1983; Behrens, 1989). Their applications are versatile, allowing for compression, neutralization, or distraction of fracture fragments, and some devices can even be adjusted in position over time to improve the fragment reduction. External fixation devices often are used for treatment of fractures associated with severe soft tissue injury or contamination of the soft tissues, or to minimize surgical trauma to soft tissues while obtaining adequate fragment fixation and stabilization. External fixation may be used in special cases, such as bone lengthening, arthrodesis, fractures requiring distraction, infected fractures, and nonunion fractures. Some types have bars that connect the pins, others have rings that partially or completely surround the limb (Ilizarov), and others have a combination of clamps, bars, and rings.

External fixation devices usually are placed so that penetration of the musculature by the fixation pins can be avoided if possible. The surface of the bone usually used for positioning an external device in the tibia is medial, for the radius craniomedial or medial, for the humerus craniolateral, and for the femur lateral.

The *components* of external fixators include the externally positioned connecting bars which are most often parallel to the shaft of the bone, fixation pins that pass through the bone at an angle, and connecting clamps that attach the connecting bar(s) with the fixation pins. All fixation pins should be within the same plane in the more usual type I and type II splintage.

In *type I* splintage, the fixation pins pass through only one skin surface and through both cortices. The fixation pins are referred to as "half-pins." Unilateral or one-half pins enter the soft tissues on one side through a small skin incision and thread directly into the bone. The connecting clamps and bars are placed on one side of the limb only. It is possible to use two connecting bars on the same side of the limb, and this is referred to as a type I double connecting bar configuration. A type I double clamp configuration also can be built with a short connecting bar attached to two one-half pins in each major fragment. The two short connecting bars are then attached to each other by clamps and a single connecting bar. This configuration generally is not recommended due to the weakness of the double clamp. It is possible to place the connecting bars cranial and lateral to the limb and use half-pin fixation, creating a triangu-

lar configuration as seen end-on. This is known as a type I quadralateral frame configuration. A type I configuration may be used to control fragment rotation in association with an intramedullary pin to control axial movement of the fragments (Fig. 4-7).

In *type II* splintage, the fixation pins pass through both skin surfaces and both bone cortices. The fixation pins are referred to as "full-pins." Full-pins enter the soft tissues on one side through a small skin incision, thread directly into the bone, pass through the soft tissues on the opposite side of the limb, and exit through the skin on the opposite side of the limb. The connecting clamps and bars are used on both sides of the limb. These are referred to as "through-and-through," "frame," or transfixation techniques. A modified type II consists of fixation pins with one full-pin in each major fracture fragment with half-pins placed for additional support (Fig. 4-8). Type II splintage usually is limited to treatment of fractures of the radius and tibia because of the requirement that connecting bars be located both medially and laterally.

In *type III* splintage, combinations of half-pin and full-pin splintage are used. A type III three-dimensional configuration is a combination of half-pin and full-pin splintage with three connecting bars, forming a triangular appearance when viewed end-on. It is known as the "tent" or "tripod" configuration. It is built with type I and type II splints placed at 90° axial rotation to each other and then interconnected at both ends, creating a three-dimensional frame (Fig. 4-9).

The type of *clamps* can be altered, with single clamps connecting a fixation pin to a connecting bar, whereas double clamps can be used to connect two connecting bars.

Pins used for fixation can be smooth or threaded, and are referred to by a variety of names, usually the name of the inventor. Steinman pins are large-caliber wires with pointed tips that are cut to a desired length and used for fixation of fractures or as traction pins. These pins may be smooth, partially threaded, fully threaded, or centrally threaded. A particular type of pin is characterized by having a raised thread located either centrally or at the tip.

Pin placement is important. It is recommended that fixation pins be placed at a 70° angle to the shaft of the bone. The clamps on the connecting bar are spaced apart, and the pins converge with

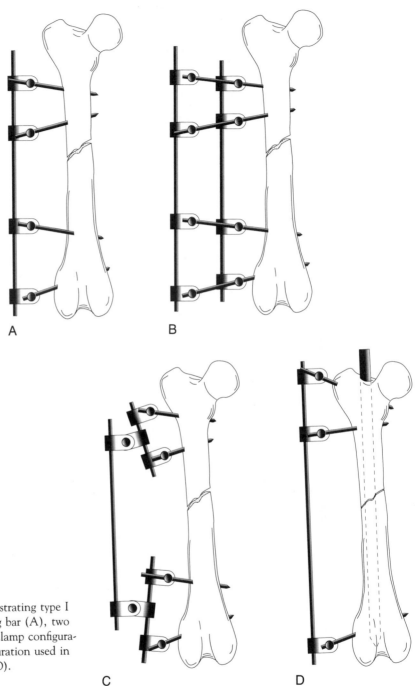

A

B

C

D

FIGURE 4-7. Drawings illustrating type I splintage with one connecting bar (A), two connecting bars (B), double clamp configuration (C), and a type I configuration used in association with an IM pin (D).

one pin in the proximal end of the fragment and one pin in the distal end of the fragment. Most fractures require two fixation pins in each major bone fragment for the most effective stabilization. In certain patients in which less support is required or when the external device is used in conjunction with other forms of fixation, only one pin may be placed in each major bone fragment. In larger patients, the use of three or four pins in each fragment provides greater support. The *number of pins* in external fixators can be used in naming the device. A 2/2 splintage uses two fixation pins in each major fragment, a 3/3 splintage uses three fixation pins in each major fragment, and so on.

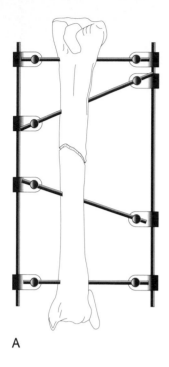

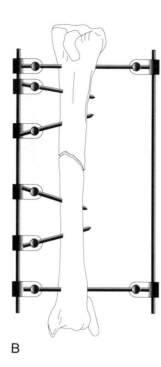

FIGURE 4-8. Drawings illustrating type II splintage, in which the fixation pins pass through both skin surfaces and both bone cortices (A). A modified type II splintage consists of fixation pins with one full-pin in each major fragment with half-pins placed for additional support (B).

A

B

Rarely is there space for placement of four pins per fragment. The length of pin between the fixation clamp and the bone should be minimized to obtain the greatest support. The fixation pin should penetrate the second cortex of the bone by a distance equal to the length of the trocar point.

Radiographic evaluation of these devices must include determination of pin placement and verification that both cortices of the bone shaft are penetrated. The entire bone needs to be included on the radiograph because the external devices may be a great distance from the fracture site. The penetration of the far tip of a pin in soft tissues should be noticed because this can be a source of pain for the patient if it protrudes into a muscle belly. Fracture of fixation pins usually occurs at the junction between the pin and bone because of cycling at this site. The interface between the threaded and smooth portions of the fixation pin is the weakest part of the pin, and this interface should be positioned within the medullary cavity and not at the pin–bone interface. This can be noted radiographically and corrected appropriately.

An external fixation is frequently used as an auxiliary function. Commonly, the external fixation device is used in conjunction with an IM pin.

In other patients, use of the external fixator is an admission of inadequate fracture stabilization, and signals that the surgeon thinks additional support is needed.

Location of fixation pins relative to other implants should be noted radiographically. For example, the proximal fixation pin placed within the humerus in association with an IM pin should be placed caudal to the IM pin, where it can be placed into a greater amount of bone tissue. If the fixation pin is positioned cranial to the IM pin, it passes through the cortex, where it causes weakening and may break out. The angle of the fixation pins needs to be noted because it determines the "grip" of the pins on the bone.

Complications in the use of external fixators are not uncommon because of their use in conjunction with complex injuries. As the fracture heals, lucency is expected around the pins as they pass through the cortices. If the lucency is of uniform size along the pin shaft; it has resulted from minimal motion of the pins and usually does not indicate an infectious process. Lucency that is unequal or patchy and occurs focally is more suggestive of osteomyelitis. Pin-track infection is a frequent complication, and is related to: (1) the technique of pin insertion, (2) care of the pin site,

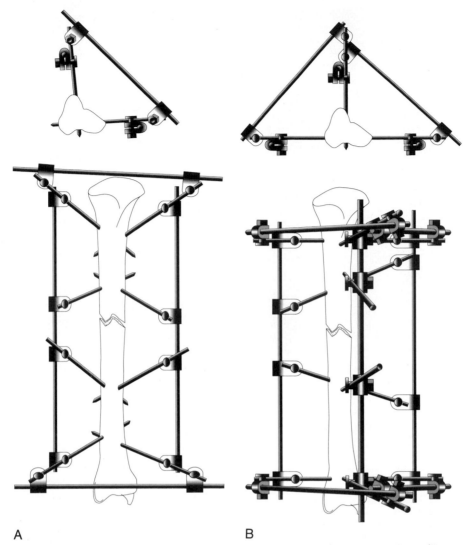

A   B

FIGURE 4-9. Drawings illustrating type III splintage, in which a "tent" or "tripod" configuration is used. Type III splintage can be constructed with type I splints (A), type II splints, or a combination (B).

and (3) stresses on the pin–bone interface. A special type of pin-track problem exists when a high-speed drill generates enough heat to result in thermal necrosis of the surrounding bone, creating what has been referred to as a "ring sequestrum." The comparative location of the pins should be evaluated on subsequent radiographs because pins may back-out due to movement or infection. The pins may bend, especially if stress is placed across the fracture site. This usually follows mistreatment of the apparatus by the patient.

Aftercare of external fixation devices is important. They should be bandaged so that the pin ends do not become entangled and loosen as a result of the patient moving within the house or yard. The bandage should be kept dry and clean to prevent pin-track infection. Often the patient requires housing in a restricted area to prevent damage to the device, and this needs to be discussed thoroughly with the client.

Intramedullary Fixation Devices

Intramedullary devices can be solid or hollow, circular, triangular, or cloverleaf in cross section, can be flexible to rigid, and can be placed in reamed or unreamed channels. Many pins are hollow and almost all have a unique cross-sectional pattern

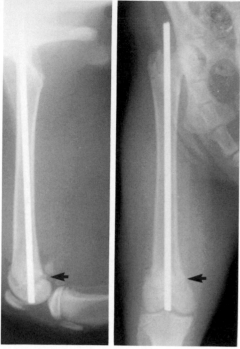

FIGURE 4-10. Lateral and craniocaudal radiographs of the femur in a kitten after fixation of a distal physeal fracture (arrows) using a single smooth IM pin. A smooth pin can be placed across a physeal plate without unacceptable risk of injury to the growth plate.

designed to reduce or prevent fragment rotation. The terms "rod," "nail," and "pin" have specific biomechanical implications, but clinically the names are used interchangeably. The devices can be used singularly, or in patterns using more than one pin (stacked). Stacking may not offer a marked increase in resistance to torsional forces (Dallman et al, 1990). It is a common practice to use an IM pin in conjunction with another fixation device (see Fig. 4-7D).

Radiographic evaluation of a fracture requires careful inspection of fragment location. The radiograph needs to be examined carefully to ascertain the presence of a long fissure fracture in addition to a more transverse fracture, because it is common for this type of fissure fracture to permit splitting of a large fragment at the time the IM pin is positioned. Detection of these additional fracture lines might discourage the use of an IM pin for fixation, or at least dictate use of cerclage wires to stabilize the potentially weakened bone. The size

of the medullary cavity needs to be accurately measured to avoid fragment splitting or separation of additional cortical fragments by forcing too large a pin into the medullary cavity.

Radiographic detection of correct IM pin placement may be difficult. Try to note the points of contact between the endosteal surface and the pin. Three-point fixation should be present, but often the selected pin is small and there are no sites of contact between pin and endosteal surface. After placement of an IM pin, the fragments of a long bone tend to telescope around the pin, with resulting shortening of the bone and additional fragmentation at the original fracture site. Radiographic evaluation of fracture healing stabilized with intramedullary devices shows a preponderance of extracortical callus because most of the medullary blood supply that would support the development of the internal callus has been destroyed by the original injury and the subsequent reaming procedure. Formation of intramedullary callus occurs slowly, filling the space between the pin and endosteal lining of the cortex, but rarely is noted on the radiograph.

Placement of IM *pins across a physeal growth plate* is acceptable only if the pin is smooth (Fig. 4-10). A threaded pin holds to the cancellous bone both proximal and distal to the plate and compresses the physeal plate, preventing bony growth or causing premature closure of the growth plate. Pins, even though they are smooth, cannot be placed in a crossing manner across a growth plate, because this also prevents separation of the metaphysis resulting in premature or delayed growth. Sometimes *pins are placed across a joint space* with the end of the pin entering the opposing bone, preventing motion of the joint. This is acceptable for a limited period of time, bearing in mind that the good health of the articular cartilage depends on continual joint motion and that periarticular soft tissue tightens with a lack of joint motion. Pins should not be placed so that the tip enters the joint space because movement of the opposing bone occurs and severe injury to the articular surface results from the gouging action of the tip of the pin. All of these possibilities can be evaluated radiographically.

Use of a threaded pin should not be considered if there is an anticipated need to countersink it beneath the articular bone. This is because the threaded tip is placed tightly within the proximal fragment, and driving the tip results in separation of fragments that were tightly reduced.

Lucency may be noted around the pins during healing. Because most long bones are not perfect tubes, the distance between the pin and the cortical endosteum at the time of placement is not constant. At certain points there is direct contact between pin and cortical endosteum, and a space is not present, whereas in other areas the space between the pin and endosteum is rather wide. Because of pin placement, identification of an infectious process developing within the medullary cavity is difficult radiographically; it is almost impossible to detect destruction within the surrounding cancellous bone after surgery. Radiolucency at points of contact within the endosteal surface could be due to motion as well as bone infection. If focal cortical osteolysis or periosteal new bone production adjacent to the suspect lesion is noted, osteomyelitis rather than simple pin motion should be suspected.

Migration of IM pins may occur during healing as a result of motion at the fracture site. The direction of migration is usually proximal. If distal migration is noted, it suggests that an original drill hole protruded through the distal cortex and articular cartilage, providing an opening through which the pin could migrate distally.

Removal of IM pins usually is carried out after fracture healing; a protruding pin tip is grasped with an appropriate instrument and retracted. If the pin is placed in a young patient, the normal increase in length of the bone may result in it being buried within the metaphysis, making removal difficult.

Radiographic Appearance of Fracture Healing

5

Importance of Blood Supply

To understand how a fracture heals, one must know the *blood supply to a bone organ*, how it has been disrupted by the fracture, how a new extraosseous blood supply participates in bone healing, and how a blood supply similar to the preinjury one becomes reestablished (Trueta, 1963; Rhinelander, 1965; Rhinelander et al, 1968). All of this determines where and when callus forms and strongly influences the radiographic appearance of fracture healing. There may be one or more principle nutrient arteries that are characteristic in number and location for a particular bone in a particular species. The epiphyseometaphyseal nutrient vessels are separated by a cartilage plate during bone growth and are then joined by anastomoses with disappearance of the growth plate at the time of skeletal maturation. The total cross-sectional area of the epiphyseometaphyseal arteries exceeds by several times that of the principal nutrient artery of the diaphysis. Experimental ligation of the diaphyseal nutrient artery has demonstrated that the epiphyseometaphyseal arteries and periosteal arteriolar network can supply adequate blood to the entire long bone. Midshaft fractures with fragment displacement demonstrate this fact frequently.

The *principle nutrient artery* of the diaphysis enters and passes through the cortex and reaches the medullary cavity without supplying collaterals to the cortex. Here it divides into ascending and descending branches that anastomose with the epiphyseometaphyseal vessels, from which branches ultimately arise to supply the bone marrow and the compact and cancellous bone. The nutrient artery passes through the nutrient foramina and often courses along the endosteal surface in a bony groove that can be identified radiographically.

Beneath a loosely attached periosteum, the full thickness of the cortex is supplied by the *endosteal blood supply* via nutrient vessels of the diaphysis. Centrifugal blood flows from the medullary cavity toward the cortical surface, where blood leaves through periosteal venules. This represents the blood flow pattern for most of the cortex of the long bone shaft of a dog or cat. In the cortex at the site of heavy fascial attachments, approximately two thirds of the inner cortex of the diaphysis is supplied by the endosteal supply (centrifugal flow), whereas the outer one third of the cortex is supplied by periosteal arterioles (centripetal flow). The blood pressure in medullary vessels (60 mm Hg) entering Volkmann's canals at the endosteal surface in the healthy dog is higher than that normally present in the periosteal vascular network (15 mm Hg).

Haversian arteries are the capillary-sized vessels of the diaphyseal cortex that are longitudinally oriented within the Haversian canals, with frequent short transverse capillaries that travel in Volkmann's canals. These horizontal vessels anastomose between the medullary and periosteal capillary beds.

Epiphyseal arteries supply the nutrient needs of the growth plate from its epiphyseal aspect (i.e., the side of the plate nearest the zone of resting cartilage). *Metaphyseal arteries* supply the rich capillary layer on the metaphyseal side of the growth plate and function not in physeal nutrition but in erosion of the hyaline cartilage of the plate at the zone of cartilage cell death. After cessation of growth with closure of the physeal plate, the epiphyseal and metaphyseal vessels anastomose. In small mammals, vessels do not cross the growth plate between the epiphysis and metaphysis before growth plate closure.

The *vascular damage* that has been sustained by the blood supply of the bone is important in the evaluation of the fracture and its expected healing (Göthman, 1961). In addition, the extent of injury to the surrounding soft tissues influences fracture healing because these tissues are the origin of the extraosseous blood supply for healing bone. It is important to attempt to determine the level of these injuries from radiographs as well as from the physical examination. As one contemplates the type of fixation to use to stabilize the fracture, it is possible to envision the further injury to the blood supply that is likely to occur during the surgical procedure. The character of the blood supply to the bone as well as to the soft tissues is a matter of paramount concern. The placement of an external K-E apparatus distant from the fracture site has little effect on the blood supply to the bone, whereas an open surgical procedure used in the placement of an IM pin or plate causes further injury to the blood supply of both soft tissues and medullary cavity. It is for this reason, in part, that a long, complicated surgical procedure to achieve perfect anatomic reduction for fracture treatment is not always in the best interests of all patients.

Bone growth can continue in the immature patient during fracture healing; in fact, it may be enhanced by an increased blood supply. This means that the location of an IM pin or fixation pin in an external K-E device moves away from the end of the bone during healing. This movement should have no effect on fracture healing or on bone growth. It does suggest, however, that it is contraindicated to: (1) place a threaded IM pin across the physeal plate, (2) place smooth pins across the physeal plate that are not parallel to each other, (3) bridge a physeal plate with a K-E apparatus, (4) bridge a physeal plate with a bone plate, or (5) place a full-threaded screw across the physeal plate, because any of these will retard bone growth either completely or partially, resulting in a shortened bone or one that grows with angulation.

Radiology of Fracture Healing by Secondary Intention

The degree of *soft tissue injury* partially determines the healing potential of the fracture. With severe soft tissue injury, the new extraperiosteal blood supply that feeds the healing fracture fails to form and delayed fracture healing or a nonunion fracture results. The nature of the soft tissue injury can be estimated from the radiograph by noting the amount of swelling and hematoma formation as well as the displacement of the fracture fragments. In addition, the presence of interposed soft tissues separating the bony fragments is an indication for potential delay in the fracture healing. Marked fragment overriding or severe comminution of the fragments are other potential indications of extensive soft tissue injury and potential delay in healing. The radiograph, at best, offers only a clue as to the extent of soft tissue injury, but should be judged as closely as possible, because, in addition to the physical examination, it often is the only evidence available for determination of this important information.

Granulation tissue leading to scar formation is the method used by most mesenchymal-derived connective tissues to repair extensive defects. However, successful fracture repair results from the production of a *callus* (a fibro-cartilagino-osseous splint), which tends to immobilize the bone fragments. If immobilization is achieved, mechanical "stress lines" develop in the healing fracture. These stress trajectories produce poorly understood bioelectrical forces that influence bone deposition and osteoclastic resorption for achieving the greatest strength for the bone while using the smallest volume of bone tissue.

Because *bone repair* depends on a recrudescent osteogenic mesenchyme and a good blood supply, the most rapid and successful repair should occur in fracture sites having abundant cancellous bone with highly vascularized red marrow (e.g., metaphyseal fractures). Fractures that heal more slowly and have more potential complications for delay are those involving compact bone with fatty marrow (e.g., mid-diaphyseal fractures), particularly those portions of bone shafts not overlain by muscle attachments, with their enhanced blood supply. Fractures in young animals heal faster and have a better chance of successful repair than fractures in older animals. Nutritional or metabolic diseases may adversely alter healing rates.

The *sequence of repair* of a closed, complete, traumatic fracture of a tubular (long) bone by secondary intention is as follows. The trauma causes hemorrhage (clot) and local blood vessel damage (periosteal, Haversian, endosteal, and often the nutrient vessel) and local necrosis of affected soft tissue and bone back to sites of intact vascular perfusion (Urist and Johnson, 1943). The significance of clot is in dispute, and opinions vary; however, most would agree that a large clot slows the healing rate.

Active hyperemia characterizes the borders of the bone injury. Although part of this is due to the release of local mediators of inflammation, at least some of the increased blood flow is the result of a "neurovascularglomic" response mediated in the spinal cord, as evidenced by simultaneous active hyperemia in the same bone of the opposite limb (Rhinelander and Baragry, 1962; Rhinelander, 1974). In this manner, the autonomic nervous system acts on local muscular arterioles to maintain a sustained, local, active hyperemia to aid in fracture healing.

Removal of necrotic tissue and bone takes place through liquefactive necrosis, with phagocytosis and osteoclasts removing dead bone. This is a slow process that continues during the course of healing and is the first radiographic sign of healing. It is recognized by fracture fragments losing their sharp borders and by a concurrent loss of bone density. The fracture site thus becomes "fuzzy" and less distinct in its appearance on the radiograph.

Proliferation of mesenchymal cells occurs along the ends of the bone fragments where they arise from the deep layer (osteogenic layer) of the periosteum and the paraskeletal soft tissues. In an immature animal, periosteal stripping removes the osteogenic cells from the surface of the bone, and they remain attached to the elevated periosteum (Whiteside et al, 1978). In mature animals, the periosteum tears instead of stripping from the bone, and the osteogenic cells remain on the cortical surface. This affects the radiographic appearance of developing callus at the fracture site with callus developing away from the bone in the immature patient and adjacent to the bone in the mature animal. "Bony callus" is the name given to the trabeculae of woven bone with some islands of cartilage present. It is visible radiographically because the osteoid has picked up enough mineral salts to become radiodense.

Mesenchymal cells also arise from both the endosteum and undifferentiated mesenchyme of the marrow. Cells from these two sources form the smaller internal callus that is not seen radiographically because it is buried within the cortices and covered by any external callus formation. All of these mesenchymal-derived cells have osteogenic potential and bridge the gap between the ends of the fractured bone.

The *periosteal component* of the callus grows more rapidly and forms a collar (external callus) around the fracture site. This initial callus, sometimes called a "fibrous" callus, appears 4 to 5 days after the fracture, but cannot be seen radiographically. It acquires enough density (bony callus) to be seen radiographically at 10 to 12 days after the injury, although the time of appearance is influenced primarily by the nature of the fracture and the age of the patient. With less severe injury to the bone and soft tissues, or in a younger patient, the callus can be seen radiographically as early as 4 to 6 days after injury.

Differentiation of the repair cell population occurs when the mesenchymal cells immediately adjacent to the surfaces of the ends of the fractured cortical bone that have a good blood supply differentiate into osteoblasts, forming trabeculae of woven bone through intramembranous ossification. Some of the bone is anchored to the bone fracture surfaces. Mesenchymal cells in the superficial portions of the external callus in the gap between the ends of the fractured bone usually differentiate into chondroblasts and lay down islands of cartilage. Reasons for cartilaginous differentiation are not clear; it is reported that it may be due to the area being ischemic, and local tissue hypoxia causes osteogenic mesenchyme to differentiate into cartilage instead of bone. This is important radiographically because the full extent of the bridging callus is not recognized radiographically if it contains a large cartilaginous portion. What is seen on the radiograph is only the ossified portion of the callus. This is part of the reason the palpable callus feels larger than the callus as seen on the radiograph. Trabecular bone continues to increase in the vascularized mesenchyme through intramembranous ossification and advances to replace the islands of cartilage by endochondral ossification.

The *initial blood supply to the external callus* is furnished by arterioles of the adjacent soft tissues, especially those of the traumatized muscles. This is the extraosseous blood supply of healing bone. By 5 weeks after fracture in a healthy dog, medulla-derived arterioles have penetrated the entire width of the cortex and have become the main blood supply to the entire callus, including the external callus.

Callus maturation takes place as the new trabecular or cancellous bone is converted into compact lamellar bone in line between the ends of bone fragments at the fracture site. Removal of necrotic bone continues. As modeling begins, the callus assumes a fusiform shape; however, the size and shape of the callus is greatly influenced by motion at the fracture site. Trabeculae peripheral to this line are gradually removed as original conformation of the bone is restored, because they are no longer needed for support. Radiographically, this is recognized as maturation of the bony callus. The density of this external mineralized mass is the safest measure of bone healing (Bick, 1948). Failure of maturation to occur suggests that healing is not completed and motion at the fracture site is delaying this important step in fracture repair.

Because stress is borne by the "tubular" cortex, the callus in the medullary cavity at the fracture site is no longer needed to provide support and begins to undergo osteoclastic resorption, allowing the reestablishment of a continuous medullary cavity in the diaphysis of the bone. Because of the size of the external callus, little of the modeling of the internal callus is appreciated radiographically. Eventually, with modeling of the external callus and its thinning, some of the change within the medullary cavity can be appreciated radiographically. Usually this is so late within the healing phase that it has little practical significance.

Motion at the fracture site is a most important feature in determination of the expected radiographic appearance of fracture healing, because motion determines the amount of callus required. Stable reduction of a fracture can result in production of little or no callus as seen radiographically, whereas a midshaft fracture with no end-to-end apposition of the fragments and treated with an external splint or cast requires a massive bridging callus to achieve healing. This callus is referred to as "the exuberant callus of fracture repair."

Radiology of Fracture Healing by Primary Intention

Direct bony union or contact healing by osseous tissue may occur between bony fragments. This form of healing occurs when reduction of the fragments results in a stable fixation with cortex in contact with fractured cortex and medullary cavity with medullary cavity. Little or no internal and external callus formation is necessary to achieve this type of healing. Almost perfect immobilization of fragments must be obtained by rigid fixa-

tion of bone fragments under longitudinal compression for this to happen. Healing and reconstruction of the cortex at the fracture line become simultaneous events. This can occur only because there is bone-to-bone contact of fracture fragments at the fracture line.

Many more *cortical remodeling units* ("cutting cone"—"closing cone" units) are activated in the cortical bone adjacent to the compressed fracture line than are present in the normal, unstressed cortex. As "cutting cones," they cross, or model, the compressed fracture line, and the "closing cone" components form new osteons that bridge the cortical fragments together with direct bony union or contact healing. "Osteonization" of the fracture line not only restores normal cortical morphology, it restores the normal vascularity through creation of the Haversian vessel pattern.

This pattern of healing can occur when a compression plate is used with or without interfragmentary screw placement. The radiographic pattern of healing is completely different from that seen in secondary intention healing. It is possible for the osteoclastic action at the fracture site to reduce the density of bone, and this is noted radiographically as a "smudging" of the interface between the fragments. With firm stabilization of the fragments, no external or internal callus need be formed because healing occurs by primary intention. This means that the most obvious radiographic change that is used to monitor fracture healing by secondary intention is not present in primary intention healing. This can be disconcerting to the surgeon who is not familiar with this type of healing and stumbles across it occasionally. There may even be a question of nonunion when healing is in fact progressing well. Use of the limb and absence of pain on palpation of the fracture site should be relied on to provide evidence of fracture healing.

Radiology of Autologous Cancellous Bone Grafts

It often is necessary to use grafts of autologous cancellous bone in orthopedic surgery. The graft is a transplant of living tissue, whereas an implant refers to nonviable material, such as metal, plastics, or dead bone, placed within the body.

Three main uses of *autologous cancellous bone grafts* are to (1) stimulate union in fractures in which one anticipates indolent healing (e.g., in patients with severely comminuted fractures, certain open fractures, fractures characterized by massive bone loss, fractures with delayed union, or nonunion fractures); (2) promote fusion (arthrodesis) of joints damaged by injury, disease, or deformity; and (3) fill bone defects due to tumors, cysts, bone loss in severe fractures, and in chronic infection. In this text, we are interested in the use of bone grafts to assist in fracture healing.

The main advantage of using an *autograft* is that the donor and the recipient are the same individual so that there is no immunologically mediated inflammatory response that might otherwise interfere with osteogenic repair. Bone healing is more rapid with autologous *cancellous* bone than with autologous *compact* bone. Compact bone is composed mostly of dense bone with a few vascular channels (Haversian and Volkmann's canals) each containing some osteogenic mesenchyme. Once the donor piece of compact bone is separated from its blood supply, all the osteocytes and most of the osteogenic cells within the vascular channels are destined to die because of the long interval required for revascularization of a piece of compact bone. Cutting compact donor bone into small pieces exposes more of the vascular channels and permits more successful revascularization of the chips when they are implanted in the recipient site. However, relatively few osteogenic cells located on the exposed surfaces of a graft of compact bone survive transplantation and are available for bone repair.

In contrast, the pieces of *cancellous* bone used in grafting have a markedly different morphology, with broad trabecular surfaces covered with numerous osteogenic cells. Large marrow spaces between the bone trabeculae contain numerous sinusoids and small-caliber vessels. When the cancellous bone is implanted into the recipient site, vascular anastomoses may occur between marrow vessels in the donor chip and the vascular bed of the recipient site, or, if that fails to happen, neovascularization of the marrow spaces in the donor chip is accomplished by vascular invasion from the recipient site. Time for vascularization of cancellous autologous grafts ranges from 7 to 14 days.

The size of the *population of living osteogenic cells* that covers the bone surfaces is important. Although the osteocytes that are buried within the trabeculae of the donor chips of cancellous bone die within a few hours after the bone is separated from the donor site, the sheet of osteogenic cells lying on many of the trabecular surfaces survives transplantation. These osteogenic cells proliferate on the surface of the implanted chips to form new bony trabeculae, which in time anastomose with new trabeculae produced by osteogenic mesenchyme native to the recipient site. Subsequent changes involve removal of the dead bone tissue of the donor graft and modeling of the bony architecture to meet the needs (stresses) placed on the bone as the fracture line heals and as the animal becomes weight bearing. Simultaneously, the bone marrow is reestablished and gradually takes on the character of either red or fatty (yellow) marrow, whichever is characteristic for that anatomic site.

Factors affecting the rate of graft incorporation include instability of the fracture, which damages the capillary bed and adversely affects the vascularity of the graft, and the age of the animal, because older animals have less osteogenic mesenchyme on bone surfaces and therefore take longer to mobilize a repair cell population. To a lesser extent, graft incorporation is influenced by: (1) the metabolic state of the animal, such as sufficient energy in the diet; (2) adequate levels and proper balance of bone minerals, vitamins, and trace elements in the diet that are needed for osteogenic repair; and (3) freedom from a metabolic disease, such as chronic renal disease, that affects normal bone maintenance.

The *radiographic appearance of the cancellous graft* and its rate of disappearance are important in the evaluation of fracture healing. If the graft is small and tightly packed among many comminuted fragments, it probably will not be recognized radiographically. If fragmentation is less prominent and the graft is larger and more loosely packed, the fracture site appears to have a dense "bony" cloud around it. The clinician, aware of the use of the graft, anticipates this appearance and it presents no frightening consequences. Because the graft remains visible during the time of expected appearance of the external callus, these two events are superimposed and the earliest callus is not fully appreciated radiographically. When the callus becomes larger and more dense than the graft, its presence can be more fully appreciated. Also, the graft decreases in density as it is resorbed, making the callus more easily recognized (Fig. 5-1).

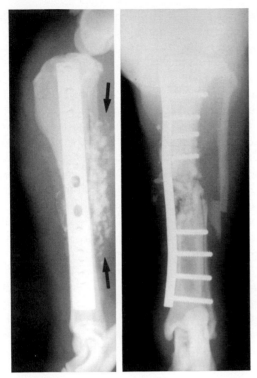

FIGURE 5-1. Radiographs of the tibia of a 5-year-old male Pit Bull with a fracture treated with a bone plate and a heavy bone graft (arrows).

Periosteal new bone is one of the early radiographic signs of osteomyelitis, and has an appearance of a thin, new bone tissue with minimal density. It is not difficult to imagine the early callus plus the cancellous graft as having the same appearance on the radiograph as an early osteomyelitis. The location of the new bone is important in determining its etiology. Healthy fracture repair centers around the fracture site, whereas bone infection causes change away from the fracture. Evaluation of the clinical appearance of the patient is important in the early days after fracture repair to determine the status of healing.

Radiology of Delayed Fracture Healing

The radiologic appearance of delayed union is exactly the same as that of normal healing except for the temporal variations. All of the features of healing are present, only in a delayed fashion. Once healing begins, one expects a progression of changes at the fracture site to continue, and de-

layed progression may not be easily appreciated as such. The surgeon is asked to make the difficult decision between a fracture that is eventually going to proceed to healing and one that requires additional treatment. The delay in healing may be caused by: (1) motion at the fracture site, (2) separation of the fracture fragments, (3) osteomyelitis, (4) the absence of soft tissues surrounding the fracture site, or (5) injury to the soft tissues surrounding the fracture site. All of these affect the blood supply available for fracture healing (Compere, 1949) or simply create a larger, potentially unstable gap between fragments over which the callus must bridge.

Often it is possible to consider the particular features of the fracture and decide that the delay in healing can be adequately explained, and that additional time can be provided for the healing. The character of the patient often assists in making the decision. If the patient is overactive physically, it may be necessary to retreat the fracture to provide additional support for a patient who is tired of waiting for healing to occur. If the patient is young, it may be justified to wait for a longer period of time for healing because of the expected presence of osteogenic cells at the fracture site. Clinically, it may be evident that callus is forming in a delayed manner and causing the patient pain because of instability at the fracture site, suggesting that immediate retreatment is necessary. So, in some patients it is possible to wait patiently without compromising eventual fracture healing, whereas with other patients it is necessary to work on the side of caution and retreat the fracture before it reaches a status where retreatment may be technically difficult or even unsuccessful. The window of opportunity during which the fracture can be successfully treated may be surprisingly small. Note that the clinical status of the patient is often as important in making a decision of whether to wait or retreat the fracture as is the nature of the radiographic findings.

Radiology of Nonunion Fractures

If fracture healing is delayed to a certain stage, signs of nonunion begin to appear on the radiograph (Sumner-Smith and Cawley, 1970; Morgan, 1972). Many of these nonunion situations are *hypertrophic*, indicating that there is a reasonable blood supply to the fracture site and all that was

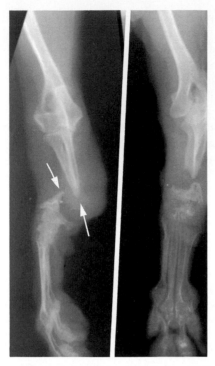

FIGURE 5-2. Radiographs of an 8-year-old Fox Terrier with a nonunion fracture of the radius and ulna in which the bone ends are not in contact with each other and disuse atrophy is occurring. Note the penciling of the fragment ends (arrows).

missing was stabilization of the fracture. The radiographic appearance of these injuries is most easily summarized by the statement that fracture healing "seems to be turned off." Nothing appears to be happening—but actually, a great deal occurs in the preparation of nonunion. The early callus formed of woven bone becomes smooth as it progresses to more mature bone, and has a nonreactive appearance. Borders of the existing callus become sharp and clearly identified compared to the less dense, immature callus at the time of its earlier formation. No active callus formation is seen. The medullary cavity at the fracture site fills with bony tissue, creating a "rounded" appearance at the end of the bone. This is called an "elephant-foot" appearance because of the expanded nature of the bone end. Bone ends adjacent to other bone ends begin to model so as to "fit together." This is indicative of the pseudoarthrosis that will develop (Fig. 5-2). Surgical debridement, including rongeuring of bone ends until they are "fresh" and

bleed, plus placement of a bone graft, prepares the site for rigid stabilization.

Nonunion may present a different radiographic appearance when fracture ends are not adjacent to each other and, instead of forming a bridging callus, the bone ends *atrophy* because of disuse and assume a tapered appearance, referred to as "penciling." Because of limb disuse, all bones become osteopenic. This is more prominently noted in small breeds of dogs, where often only a shell of bone is seen on the radiograph (see Fig. 5-2). Treatment can be attempted after freshing of the fragments ends. However, the bone atrophy is difficult to overcome and many of these treatments, even though they include rigid stabilization and grafting, are unsuccessful.

A third form of nonunion takes place when *infection* is present (Fig. 5-3). The nonunion site needs to be debrided and sequestra removed, and the fracture site requires irrigation after which a massive cancellous graft can be put into position.

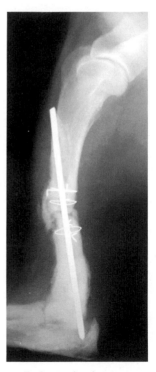

FIGURE 5-3. Radiograph of a nonunion resulting from an osteomyelitis preventing the callus bridging the fracture.

Plating to achieve rigid fixation plus the addition of a bone graft often results in fracture healing.

Radiology of Posttraumatic Aseptic Necrosis

Posttraumatic aseptic necrosis is a unique lesion seen in anatomic sites in which it is possible for a fragment of bone to be isolated and its blood supply destroyed at the time of the injury. Thus, it is another manifestation of infarction of the bone. Generally, it is the result of laceration, thrombosis, embolus, or endarteritis affecting the blood vessels, which usually are epiphyseal. This type of lesion occurs most commonly in the capital epiphysis of the femur as a result of interruption of the major blood supply, which is extracapsular and therefore highly vulnerable to trauma. With physeal separation in the skeletally immature patient or a high intra-articular femoral neck fracture in the skeletally mature patient, the potential or real blood supply through the femoral neck is broken,

and the blood supply that courses over the surface of the femoral neck within the soft tissues may be interrupted as well. The femoral head and possibly a segment of the proximal femoral neck are left without sufficient blood supply; that passing through the ligament of the femoral head supplies only the bone tissue immediately around this ligamentous attachment. Because of this problem in blood supply, intra-articular femoral neck fractures or slippage of the capital epiphysis are important clinically, and healing may not be satisfactory even with stable fragment reduction (Fig. 5-4). Occasionally, the fracture site is just distal to the capsular attachment, and the blood supply to the capital epiphysis is maintained and a more nearly normal healing occurs. The humeral head also is intracapsular, but aseptic necrosis after trauma often is not recognized in this bone of the dog and cat. Carpal and tarsal bones probably have unique solitary blood supplies that, after fracture, could result in an avascular fragment. Although this type of lesion is common in humans, it is uncommon in the dog or cat.

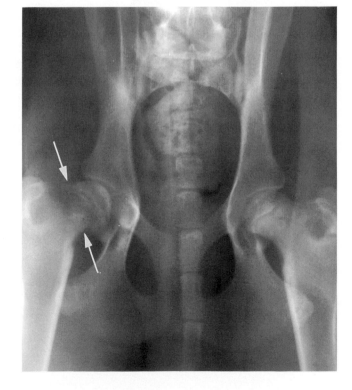

FIGURE 5-4. Radiograph showing post-traumatic aseptic necrosis to the femoral head after slippage of the capital epiphysis. Note how the avascular fragment retains its density because of lack of a blood supply, whereas the femoral neck, which maintains its vascular supply, undergoes marked osteoclastic change because of disuse and assumes the pattern of an "apple core" (arrows).

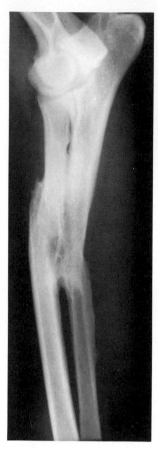

FIGURE 5-5. Lateral radiograph of the antebrachium of a dog showing malunion of radial and ulnar fractures with shortening of the bones and caudal angulation of the distal fragments. Note the synostosis between the two bones.

Radiology of Posttraumatic Osteopenia

Bone atrophy often is associated with apparently minimal bone injury, and is called posttraumatic osteoporosis, or Sudek's atrophy, in humans. Although this decalcification of bone is well characterized in humans, it is incompletely described in animals. It is assumed to be a vasomotor response initiated by pain impulses. The bones of the feet appear to be the most common site of involvement in the dog, with cloudy or patchy osteoporosis progressing to a picture of reduced, rarefied trabeculae, and cortical thinning leading to complete disappearance of the bone.

Radiology of Malunion Fractures

Malunion is the joining together of fracture ends producing a bone organ that is unacceptable because of: (1) bone shortening, (2) angulation of the distal fragment, (3) rotation of the proximal or distal fragment, or (4) osteosynthesis between adjacent bones. The pattern of fracture healing is normal and histologic examination of the bridging callus or the resulting modeled bone is normal. Only the bone organ is abnormal, and the manner in which the limb can be used is unacceptable (Fig. 5-5). Treatment may consist of refracture with possible wedge osteotomy, rotational osteotomy, or use of a bone-lengthening procedure. Usually the surrounding soft tissues are healthy and the resulting refracture heals, assuming rigid fixation of the fragments is obtained.

Radiology of Osteomyelitis

6

Bone infection, *osteomyelitis*, may be seen anywhere within the fractured bone or specifically at the site of fracture healing. Although the term suggests an inflammation of the bones (osteo) and medullary cavity (myelitis), in common usage, osteomyelitis refers to an inflammatory, infectious process of any or all bone tissue. Because of the intimate association of bone and bone marrow, both tissues are commonly involved in an inflammatory process occurring throughout the bone organ. It usually is impossible, radiographically or through clinical or gross postmortem examination, to determine the precise site and extent of the infection in the bone without histologic confirmation. Actually, the term *osteitis* might be better nomenclature in patients with bone infection because changes within the dense cortex are responsible for the major change noted on the radiograph, and disease within the marrow cavity affecting the sparse trabecular bone is almost impossible to identify radiographically.

Pathogenesis of Osteomyelitis

An understanding of the pathogenesis of osteomyelitis may help in an understanding of how certain fractures become infected and how the infection progresses and is finally brought under control. A major part of this understanding depends on knowledge of the *blood supply* to the healthy, nontraumatized bone, and the alterations that occur with trauma and during healing of the fracture and injured soft tissues. If the blood supply to a bone or fragment of bone is compromised, the possibility of successfully combating an infection is greatly limited and an osteomyelitis is more likely to occur. When injury to the blood supply occurs, stasis of blood flow and pooling of blood creates excellent culture sites.

Disruption of the blood supply occurs in a different manner in each fracture. If the fracture is midshaft, it destroys the origin of flow through the

nutrient vessels and causes a greater reliance on the epiphyseal–metaphyseal complex. If a butterfly fragment separates with an intact muscle tendinous attachment, a persistent blood supply to that fragment comes through the soft tissue attachment. However, if the tendon attachment is torn or the fragment has no tendinous attachment, the fragment becomes avascular and a new blood supply to that fragment must come from an extraosseous source originating from the surrounding soft tissues. In a badly comminuted fracture, most of the small comminuted bony fragments are without a blood supply. Continued trauma to the fractured limb after the fracture and before stabilization affects the character of the soft tissues, further compromising revascularization and healing of the fractured bone. An avascular fragment is not synonymous with osteomyelitis. The fragment may develop a new blood supply, if it is within a friendly soft tissue environment.

The *method of fracture reduction and stabilization* influences the patient's ability to initiate a new blood supply to the fractured fragment. If the marrow cavity is reamed to permit placement of IM pins, it is obvious that many remaining medullary vessels that were intact will be destroyed by the surgical procedure. In an effort to permit space for regrowth of the medullary vessels, use of fluted pins has been touted rather than the use of round pins, which fill the marrow cavity more completely. Overly aggressive repositioning of fracture fragments or tearing of muscle tendon attachments in an effort to position a bone plate or bone screws influences soft tissue viability and vascular regrowth from the soft tissues. Placement of a bone plate prevents the ingress of vessels from the soft tissues to a limited area of the bone surface beneath the plate. Placement of a cerclage wire that ultimately slips results in the wire "amputating" the developing capillary buds as they try to enter the bone.

The *osteomyelitis may be located* within a portion of the bone that is surrounded by relatively healthy soft tissues, albeit traumatized, and thus has an intact blood supply. This fractured bone is therefore susceptible to infection in a manner similar to any bone after an injury such as a bite wound causing a soft tissue infection, a hematogenous infection of the bone, or direct implantation of the microorganism within the bone. The conditions required for the development of the osteomyelitis are the same, that is, thrombosis of the vascular supply with subsequent colonization of the microorganism. This form of osteomyelitis might be present for some time without radiographic visualization of a sequestrum, or the infection may be within cancellous bone tissue in which sequestration rarely occurs.

The *experimental production of osteomyelitis* through vascular injection of microbial organisms alone is unsuccessful, and additional conditions within the tissues are required. Injection of *Staphylococcus aureus* only or sodium morrhuate (a sclerosing agent) only in rabbits, produced no gross or radiologic evidence of osteomyelitis, and bone cultures obtained at 14 and 50 days after injection were consistently sterile. However, an injection of a combination of both bacteria and sclerosing agent induced osteomyelitis (Norden, 1970). These failures to induce osteomyelitis are thought to be due to the requirement of bacteria plus vascular thrombosis. The thromboses might result from either: (1) blood clot, (2) plasma extravasation, (3) stasis of blood flow, or (4) tissue necrosis. Vascular stasis or occlusion causes plasma transudation, which favors the localization of bacteria and provides culture media for growth and colonization. It is not difficult to imagine how the trauma that caused the fracture can provide these additional requirements for the development of an osteomyelitis.

An understanding of the *pathogenesis* of bone infection makes it clear why it is almost impossible to have sequestration of cancellous bone. The trabeculae are thin and present many surfaces, so it is difficult to destroy the blood supply to an area of this kind of bone. The surface area-to-mass ratio of cortical bone is 0.2 cm^2/g, whereas that of trabecular bone is 9.0 cm^2/g (Kahn and Pritzker, 1973). Understanding the pathogenesis of infection also helps to explain the susceptibility of immature bone to cortical sequestration. Thrombosis of blood flow within the center of the bone is one

important event. Next, the periosteal blood supply is destroyed because of the looseness of the attachment of the periosteum that permits its elevation by exudative fluids or by stripping at the time of the trauma. In this way, the internal blood supply to the cortex as well as the blood supply from the surrounding soft tissues are destroyed. It is thus relatively easy, by destroying the blood supply from both the periosteal and endosteal sources, to create an avascular cortical fragment that becomes a sequestrum in the presence of soft tissue infection.

The *agent of infection* may be bacterial or it may be fungal. If the infection is bacterial, it may be due to anaerobic or aerobic agents. The infection may be purulent or nonpurulent. Generally, osteomyelitis associated with fractures and fracture healing in small animals is bacterial, aerobic or anaerobic, and often purulent.

Once the bacteria gain entrance to the fracture site, they adhere to damaged endothelia of blood vessels where the basement membrane is exposed, proliferate, and provoke an inflammatory response. The colony then increases in size if conditions are acceptable. Unless the colony is overwhelmed by the initial inflammatory response, a new focus of infection is established, resulting in an expanding area of inflammation. It is characterized histologically by those changes associated with any acute purulent inflammatory process, that is, edema, hyperemia, hemorrhage, exudation of neutrophils, and necrosis of bone and soft tissue of the bone marrow. Subsequent bone resorption is led by osteoclasts. The exudate in *acute osteomyelitis* may be under pressure because of the incompressibility of the surrounding bone tissue. Osteoclasts enlarge the walls of the Haversian and Volksmann's canals, permitting extension of the septic exudate, spread of infection, and drainage of the lesion (Harris and Kirkaldy-Willis, 1965). The *development of the osteomyelitis*, as with infection in general, depends on an intimate interplay between the microorganism and the host. Features influencing this interaction include the: (1) anatomical properties of the involved bone, (2) reparative power of the tissues, (3) cellular and humoral immune mechanisms, and (4) virulence of the organism.

Organisms such as *Staphylococcus, Streptococcus, Corynebacterium, Salmonella, Proteus, Pasteurella,* and *Pseudomonas* species, as well as *Escheri-*

chia coli often cause suppurative bone infection. *S. aureus* is the organism most frequently recovered from both animals and humans with osteomyelitis (Waldvogel et al, 1970; Caywood et al, 1978; Hirsh and Smith, 1978; Walker et al, 1983).

Chronic osteomyelitis produces a nonsuppurative exudate composed mostly of macrophages, lymphocytes, and plasma cells. Pockets of suppurative inflammation also may be present, creating in some lesions the appearance of chronic-active inflammation. Granulation tissue attempts to organize the exudate in much the same manner that takes place in the resolution of inflammation within the soft tissues. Extensive amounts of reactive bone and fibrous tissue are formed within and on the surfaces of chronically inflamed bone organs. As a result of this fibrotic response many areas within the bone organ have a compromised vascularity, making it difficult for systemic antibiotic therapy to reach sequestered microbial organisms.

The term *epiphysitis*, when taken literally, is limited to an infection localized to a secondary ossification center and is used to describe lesions most often in the young animal in which skeletal maturation has not yet occurred. Osteomyelitis of the epiphyses may result from the original fracture. It can also occur as a result of placement of fixation pins, or immediately or later during fracture healing through hematogenous spread. "Epiphysitis" frequently has been used incorrectly in the past to describe an inflammation centered on the growth plate or physis of a long bone. This is more correctly called a "physitis."

The term *physitis* is used in reference to infection localized to a primary growth plate within long bones, but also may be used relative to infection within vertebral body growth plates. It also can be used in reference to growth plates of apophyseal centers; however, this is better referred to as an "apophysitis." Physitis, when present, would be expected to have a negative effect on longitudinal growth of the bone, causing either cessation of growth, delayed growth, or unequal growth throughout the physis.

The location of the osteomyelitis may be influenced by many factors. The infection may be localized to the external surface of the bone and provoke a periosteal reaction, *periostitis*, without substantially involving the cortex or medullary cavity. An example of this is involvement of the bone surface by extension of the infectious process from surrounding soft tissues to the fracture site. The term *osteoperiostitis* is used occasionally to describe an infectious periostitis that has originated from the periosteum and has come substantially to involve the underlying fracture fragments at a later time. The fracture fragments may have a blood supply and all or a part of this vascularized bone becomes infected, or the infection can be soft tissue-oriented and surround an avascular bone fragment, causing development of a *sequestrum*. In another patient, the infection can be associated with metallic stabilization devices and result in lucent cavities around the pins or screws, causing them to become loose. It is important to understand that bone infection associated with fractures and their healing can present several different pictures clinically and radiographically.

Bone infection influences fracture healing by causing minimal or marked differences in the rate and type of callus formation, and produces subtle or marked changes in the radiographic picture. Bone infection associated with fractures and fracture healing occurs in several ways, with variable clinical significances. In some fractures, healing is markedly affected by the bone infection and does not occur normally, ending in greatly delayed healing, malunion, or nonunion. In this group are fractures that become infected, become unstable, and require additional complicated surgical procedures to debride the wound and establish new forms of stabilization. In other fractures, in which the healing is not progressing as quickly as expected, osteomyelitis is suspected and the pattern of healing must be controlled carefully, with radiography as one method for determining prognosis and the need for therapy. From this last category, some fractures finally heal in a satisfactory manner without any additional therapy, while others require surgery because the infection is severely hindering the healing process. It is probably safer to overtreat these worrisome patients because the result of a nonunion osteomyelitis that is not provided with surgical restabilization when needed is often amputation.

The use of *antibiotic therapy* may be instituted in many patients with healing fractures that are suspected of being infected. Many of these fractures have the blood supply to the bone compromised, and the use of systemic antibiotics may control the infection within the soft tissues, but often is ineffective in controlling the infection in the bone. The fibrotic response associated with formation of granulation tissue causes many areas within the bone organ to have poor vascularity, thus making it difficult for systemic antibiotic therapy to reach sequestered microbial organisms.

The time the infection makes its appearance may vary widely. The fracture may be initially infected: (1) at the time of injury, (2) at the time of surgical repair, or (3) later, during the time of fracture healing. An infection present at or near the time of the original injury greatly influences early callus formation, may lead to instability, and may be difficult to recognize radiographically. Infection occurring later, such as in association with fixation pins, may be away from the fracture site, of little significance clinically, and not be noted until the later stages of fracture healing.

Routes of Infection

The *infectious agent gains access* to the fractured bone in a variety of ways, and may: (1) invade as a direct result of the trauma itself, with seeding occurring through an open fracture site; (2) spread from infected soft tissues adjacent to the fracture site; (3) invade through a break in aseptic surgical technique; or (4) gain access through a hematogenous route. All of these situations can lead to osteomyelitis and result in delay or prevention of fracture healing. The infection may be the result of the injury or it can be iatrogenic in nature. The infection may center at the fracture site, involve another part of the bone affected by treatment, such as an area of pin placement, or involve a part of the bone without relation to either the fracture site or an area associated with reduction or stabilization.

Hematogenous localization of pathogenic microbial agents occurs in bone organs during a bacteremic–septicemic phase of a systemic disease that often is subclinical. The original site of infection may: (1) relate directly to the traumatic event that caused the fracture, such as a deep muscle wound; (2) depend on a distantly located soft tissue infection that became infected at the time of trauma, such as pneumonia secondary to lung trauma; or (3) be a preexisting infectious problem unrelated to the trauma, such as a chronic prostatitis or chronic periodontal disease. After a bone

fracture, the damaged soft tissues adjacent to the fracture site offer an ideal culture medium and become predisposed to bacterial colonization, especially when the blood supply may be compromised. Soft tissue infection prevents the development of the new extraosseous blood supply needed to heal the fracture, and often separates nonviable bone fragments that then develop into sequestra. Certainly, where there is injury to the bone and its environment, the possibility of hematogenous spread of infectious agents is greatly increased (Waldvogel et al, 1970).

Direct implantation of the infectious agent can occur *at the time of the injury* through a puncture wound causing a break in the skin. An open fracture may be created at the time of the injury, or may be self-imposed by the patient as it abuses the limb after the initial trauma. Infection may occur secondary to a gunshot wound, although not as commonly as might be expected. In addition, *iatrogenic implantation of the microorganism* may take place during surgical stabilization of the fracture. There must be sufficient injury of the tissue, leading to vascular thrombosis, to affect the blood supply and permit establishment of the infectious process. Often the same strain of an organism can be repeatedly cultured from orthopedic patients treated within a particular surgery, suggesting a nosocomial problem.

A *contiguous focus of infection* can spread to an adjacent fractured bone, causing an osteomyelitis. The most common pattern of spread occurs in fractures adjacent to a tooth with an infectious periodontitis or an infected paranasal sinusitis (Waldvogel et al, 1970). These preexisting foci of infection are unrelated to the trauma that caused the fracture. Other than these examples, this type of osteomyelitis within a fracture site is uncommon.

Influence of Soft Tissue Injury

The character of the *soft tissue injury* around the fracture influences the onset of bone infection and how it may develop. The degree of soft tissue injury may be influenced by the severity of the original trauma, ranging from minimal injury in a low-energy fracture in a puppy whose foot was stepped on, to a more severe category that occurs in association with a degloving injury in a patient dragged behind a pickup truck. Variations in the treatment given the fractured limb by the patient between the time of trauma and the time of fracture treatment are important; additional, extensive soft tissue injury is all too common during this interval. Additional soft tissue injury also may be superimposed onto the originally traumatized soft tissue due to a lengthy or rough reduction and stabilization. Additional soft tissue injury also may be caused by motion at the fracture site secondary to inadequate stabilization. Subsequent injury to the soft tissues during the time of fracture repair may result from a noncooperative patient. Although some of the soft tissue injury occurs outside of the surgeon's control, soft tissue trauma at the time of surgical repair and the effectiveness of the stabilization device are at least partially within the control of the surgeon.

Fracture treatment is always a question of balancing whether it is better to leave the fracture without perfect reduction and expect a delayed healing that is less than anatomic, or whether it is better to achieve perfect anatomic reduction through an 8-hour surgery in which soft tissue trauma is necessarily extensive (rough), and later discover that the soft tissue bed is readily available for colonization of hematogenous bacteria. Reports suggest that most cases of osteomyelitis originate from the original trauma (Caywood et al, 1978; Walker et al, 1983). This is questioned by the authors, who have noted that a great many patients in whom osteomyelitis developed underwent operative procedures that were of great length or were carried out by inexperienced surgeons.

Some fractures with associated severe soft tissue infection are best treated originally, and for only a short period of time, with a Robert Jones bandage until the soft tissue infection is controlled, before attempting repair of the fracture.

Clinical Diagnosis of Osteomyelitis

The *clinical signs* associated with an infected fracture in a healing phase often are difficult to separate from the clinical signs associated with the healing fracture itself. In some patients, pain is recognized as it reappears within the fractured limb at a time when the patient is beginning to show signs of recovery and use the limb after fracture stabilization. This sign is often characterized by a sudden failure to use the limb after having begun tenderly to use the limb to a greater extent with each passing day. Tenderness, heat, swelling, and redness also may be present but may be difficult to identify because of the soft tissue injury associated with the fracture or subsequent reparative surgery. Fever is uncommon and not a frequently noted sign. A draining tract may call attention to the possibility of infection, especially if fixation pins have been used. However, this type of infection often is limited just to the pin tracks and may not affect fracture healing.

The *stage of the osteomyelitis* may influence clinical signs. The osteomyelitis may be acute, subacute, or chronic, and may pass from one stage to another or remain quiescent. There also may be acute exacerbation of a chronic process, suggesting an indolent disease. If antibiotics were administered indiscriminately during the early phase of fracture repair, there may be a fever of unknown origin and a masking of other clinical signs at a later time. Often the infection is chronic, partially controlled by antibiotics administered in a random manner, and not always evident clinically.

A *definitive diagnosis of osteomyelitis* may be made only by biopsy and culture of a tissue specimen taken from the fracture site, culture of drainage from soft tissue tracts, or blood culture. Examination should be made for both aerobic and anaerobic bacteria. However, culture of fluid from a draining tract often yields bacteria associated with the environment, and these usually are not the ones causing the osteomyelitis (Lewis et al, 1978). Specimens selected for culturing at the time of surgery should consist of isolated sequestra, intact bone or periosteum suspected of being infected, or soft tissues immediately adjacent to bone believed to be infected (Walker et al, 1983). Samples submitted for culture also should be assayed for sensitivity. In many patients, a *tentative diagnosis of osteomyelitis* is made subjectively simply by noting a patient undergoing satisfactory progress after fracture repair and stabilization who suddenly follows a different clinical course and refuses to bear weight on the injured bone.

Pathogenesis of Sequestration

In the center of the fracture site it is possible to find an avascular fragment of bone that has been broken from a major fragment. If this bony fragment is small and is positioned within a *healthy soft tissue environment*, a new blood supply will invade the area and the fragment ultimately will be removed by osteoclastic activity or be incorporated within the healing fracture callus. If the fragment is large or separated from the fracture site, it may persist relatively unchanged for a period of time before resorption. It is this fragment that is difficult to interpret correctly when seen radiographically.

When the fragment is avascular and located within an *unhealthy soft tissue environment* that cannot provide neovascularization, it is isolated from a blood supply and may be surrounded by pus and granulation tissue. In this situation, the fragment is referred to as a sequestrum. The fragment often comes to harbor bacteria and thus can serve as a potential source of reinfection, because antibiotics or antibodies cannot penetrate into the dead bone.

The *sequestrum* identified after a fracture can assume several forms and either be: (1) an avascular bone fragment separated by the trauma and unable to revascularize; (2) a vascularized bone fragment that becomes avascular due to vascular thrombosis and death of osteocytes after the trauma; or (3) a portion of a tubular bone attached to viable bone that undergoes vascular thrombosis after the trauma. A sequestrum associated with fractures most frequently develops from a cortical bone fragment, often large enough to be referred to as a "butterfly fragment." The larger and more dense the bony fragment, the more difficult it is for the body to initiate neovascularization and resorb it or incorporate it into the healing callus.

In a sequestrum within the center of a comminuted fracture, the development of an *involucrum* cannot be separated from the surrounding callus that is forming a bridge around the fracture site. Thus, although an involucrum with a cloaca is expected to develop and thus serves as a diagnostic aid in identification of osteomyelitis of a nonfractured bone, identification of an involucrum with a cloaca is not expected in osteomyelitis at a fracture site. However, it may be possible for the osteomyelitis to be located away from the fracture site because of pin placement or through the spread of a soft tissue infection. This creates a situation similar to an osteomyelitis in a nonfrac-

tured bone, and an involucrum with a cloaca may be identified. Later in a malunion fracture, a type of involucrum may be identified as the callus matures and surrounds the sequestrum. One situation in which an involucrum might be more clearly identified is in association with a pattern of osteomyelitis surrounding a fixation pin, where the involucrum is recognized by the sclerotic rim of new bone that forms around the lucent zone next to the pin. In the event a portion of a large bony fragment dies, a lack of any periosteal response from the dead bone tissue is seen, in marked contrast to the collar of new bone forming from the viable bone at the interface. This may be considered a form of involucrum formation.

An opening—a *cloaca*, or sinus—may form within the involucrum surrounding the sequestrum. Exudate, bone debris, and even the sequestrae eventually may drain through the sinus track(s). The cloaca usually is not well identified within a recently fractured bone. As the fracture continues to heal around the sequestrum, the cloaca may become more easily identified. Occasionally, it is possible to see something resembling a cloaca in association with osteomyelitis around a fixation pin.

An avascular fragment in a healthy soft tissue environment regains a blood supply and eventually becomes a part of the healing callus. With the new blood supply, it can model and eventually fit into the pattern of the new, healed bone. It is its position as an avascular fragment within an infected pocket of soft tissue that results in its failure to revascularize and subsequent infection. Thus, the presence of a bony fragment alone does not determine whether it becomes a sequestrum. In fact, any intact persistent external blood supply to the bone through a muscle tendon attachment along with the healthy character of the surrounding environment should ensure its renewed or continued viability and joining into the fracture healing. Even if unattached, bone fragments should not be "thrown away" at the time of surgery because these avascular fragments are incorporated into the callus later, if the soft tissue bed is healthy.

Movement at the fracture site also is important in determining if a bony fragment becomes a sequestrum. In the presence of continued fragment motion and an environment with some compromise in the character of the soft tissue, any new capillary bed finds itself being "cut off" by the movement, and the bone fragment does not revascularize.

It also is possible for a sequestrum to form on the end of a tubular bone fragment, with the dead bone remaining physically attached to the viable bone. This is not commonly found in a fracture situation; however, it is rather easy to recognize radiographically because of the absence of any periosteal new bone formation on the sequestered portion.

The presence of a good extracortical soft tissue blood supply rather than the nature of the fracture may be more important in determining the frequency of bone infection after a fracture. For example, the distal radius–ulna and the distal tibia are areas that seem to be involved frequently with osteomyelitis after a fracture, and yet severe midshaft femoral fractures rarely become infected, even though the degree of bone injury is extensive.

Histologically, the sequestrum at the time of its formation has a central zone of necrotic bone tissue, fibrin deposits, and massive polymorphonuclear cell infiltration. The surrounding area has few signs of acute inflammation but contains granulation tissue, lymphocytes, and plasmacytes. The involucrum is made up of fibrous tissue surrounded by new bone (callus), with lamellar structure depending on the age of the lesion.

The *future of the sequestrum* varies. The repair process is due to eventual hyperemia, with granulation tissue from viable tissue bringing the new blood supply to the bone. Fibrosis leads to formation of fiber bone and eventually mature bone. Thus, resorption of the sequestrum depends on contact of the bone with viable granulation tissue. If this occurs, and it may be delayed for months or years, demineralization and matrix removal follow. Sequential radiographs show the slow disappearance of the sequestrum; however, if the granulation tissue plus surrounding purulent fluid keep a budding capillary bed from reaching the sequestrum, it may remain static for years, causing persistent discharge through sinus tracks. Sequential radiographs show the fragment remaining and a pattern of new bone encompassing the sequestrum (a callus turned into an involucrum). Slow resorption reflects the ability of a new capillary bed to

obtain a partial ingrowth into the avascular fragment, but in a very delayed fashion. The greater the delay in the fragment resorption, the more likely the surrounding bone is to achieve a mature inactive status. In this event, with final disappearance of the sequestrum, a large cavity remains bordered by smooth mature bone. This is clearly seen radiographically.

So, what is the *role of infection in the creation of a sequestrum?* The infection can remain entirely within the soft tissues and cause an avascular fragment to become a sequestrum. The infection can be within the bone at the fracture site and cause thrombosis and bacterial colonization, resulting in sequestration of a fractured fragment or of the end of a major fragment. The infection can be within the bone away from the fracture site, with the sequestration occurring as it would in a nonfractured bone.

Radiology of Osteomyelitis Associated with Fracture

The *role of radiographic examination* in the diagnosis of bone infection associated with a fracture is not always clear. Radiographic changes are not evident early in the course of bone infection, so the possibility of a false-negative evaluation is very real. The radiologic study is just a momentary glimpse of a continuing pathologic process, and the evaluation of radiographs must be made with this in mind (Butt, 1973).

Local destruction of bone is reflected radiologically by lytic lesions (Waldvogel et al, 1970) and occurs along with reactive new bone. Others have suggested that bone lysis is more common in early infections, and sclerosis and sequestration are found far more commonly in the more chronic infections. The nature of the destructive portion of the lesion is influenced greatly by the localization of the lesion. With rapid spread throughout the bone, there will be a patchy lysis of the bone that is difficult to visualize through the developing callus, whereas a localized lesion creates a larger lytic zone (Figs. 6-1, 6-2).

Periosteal response is seen with bone infection, but is especially difficult to appreciate in the face of developing callus. One helpful difference is that callus forms at the fracture site and seems to be meaningful and orderly in its formation, whereas

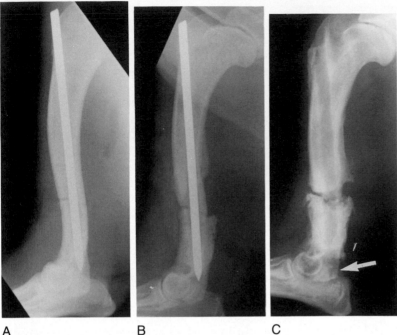

A   B   C

FIGURE 6-1. Lateral radiographs of the humerus of a 2-year-old male mixed breed dog that developed osteomyelitis after IM pinning of a fracture. A severe laceration was present at the time of the original injury that extended into the axilla (A). The IM pin was noted to be loose at 17 days after surgery, and periosteal new bone was forming away from the fracture site (B). By 3 months, the pin had migrated proximally and fallen out. The nonunion fracture with periosteal response is typical for an osteomyelitis (C). Note the destruction of the olecronon process adjacent to the elbow joint (arrow), as the infectious process spread from the humerus across the joint space into the ulna.

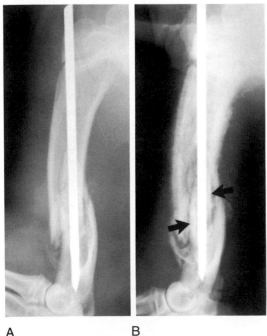

FIGURE 6-2. Lateral radiographs of the humerus of a 9-month-old Great Dane made at the time of injury (A) and 1 month after placement of the IM pin (B). The sequestrum (arrows) is easily identified through the surrounding radiolucent zone. Note the prominent periosteal response throughout the length of the bone generated by the infectious process. This is a typical pattern of radiographic changes associated with an osteomyelitis.

A   B

the periosteal response associated with the infectious process may occur through a greater length of the bone than expected and appears to have little purpose relative to bridging the fracture site. If the periosteal attachment is loose, especially in the immature animal, the collection of purulent materials allows elevation of the periosteum and formation of a new surface to the bone with lamellar patterns (see Figs. 6-1, 6-2). In the adult animal, periosteal response arises from the attached periosteum and assumes the form of "fingers" of new bone perpendicular to the bone surface.

Soft tissue changes associated with the bone infection are superimposed over the soft tissue trauma associated with the injury, and are of little value in the detection of bone infection in the face of a healing fracture.

The *time the osteomyelitis begins* influences the radiographic appearance. If the infection begins at the time of the trauma or of fracture stabilization, the early bone changes associated with the infection occur at the time of the appearance of the early bone callus. This all occurs at approximately 7 to 10 days. The first changes of osteomyelitis are characterized by bone lysis with cortical lucency, called "tunneling," or endosteal resorption, referred to as "scalloping," and can be confused with the normal minimal bone resorption at the fracture site and the early laydown of callus associated with fracture healing. If the infection begins later during fracture healing, it is possible for all changes to be lost radiographically within the healing callus. If rigid stabilization of the fracture fragments is achieved, minimal callus is required and the changes of bone infection can be seen more easily on the radiograph.

Sequestrum formation may be seen with fragments of cortical bone involved. The sequestrum retains its density as well as its sharp borders because there is no viability and thus no cellular activity that can create any change in density, size, or shape (see Fig. 6-2). In normal fracture healing, it is expected that the bony changes of resorption and early callus deposition should involve all of the bone tissue at the same time and in the same manner. However, this expectation is rarely borne out because of the variation in blood supply to the fragments. Therefore, in a healthy fracture environment, a single avascular fragment may retain normal density and have the appearance of a sequestrum, when in fact there is only a

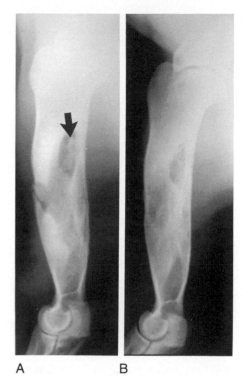

A   B

FIGURE 6-3. Lateral radiographs of the humerus in a young Pit Bull after a traumatic injury. The first study (A) was made after healing of the oblique fracture around the sequestrum (arrow). A second study was made 2 years later after the sequestrum had resorbed, leaving the involucrum. Note that the involucrum is modeling and becoming smaller, and the cloaca through the cranial cortex has closed.

delay in revascularization of a healthy fragment of bone.

If *bone infection centers around metallic devices*, the smooth surfaces inhibit fibrogenesis and thus retard the localization of infection (Kahn and Pritzker, 1973). The resulting radiolucency is uneven in width and spreads widely. The edge of the lucent zone is less sharply defined than that noted in the healthy bone-to-metal interface. Lucency due to motion can be identified around pins, creating a pattern that is uniform in development and often bordered by rather dense bone. This pattern of lucency is predictable when consideration is given to the nature of the motion of the pins.

Fracture healing may occur around the persistent osteomyelitis that might include a sequestrum (Fig. 6-3). It is also possible for the osteomyelitis to subside in the face of a delayed-union fracture.

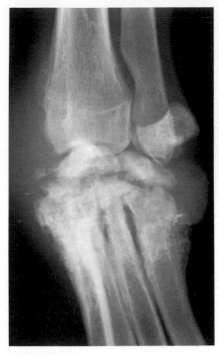

FIGURE 6-4. Radiograph of the carpal region of a dog. The injury was originally centered on the metacarpal bones, but infection quickly spread to the adjacent joints, causing a severely destructive arthritis. The bone injury obviously became of minimal clinical significance as the joint disease developed.

These cases are impossible to interpret radiographically, and it is important to remember that many patients with worsening radiographic signs of osteomyelitis actually are improving clinically (Waldvogel et al, 1970). This is partially due to the delay in bony changes as detected on the radiograph. The clinical status of cases of bone infection must be carefully monitored, and the radiographic changes used principally to detect a deviation from the usual progression of changes.

Surgical intervention may be required in a patient with an infected fracture site. The timing of secondary surgery is difficult to determine, but it should be performed sooner instead of waiting until clinical and radiographic evidence of disaster is evident. Good restabilization of a fracture should be attempted in the face of osteomyelitis.

Complications of Osteomyelitis

The most common complication of osteomyelitis is spread of the infection into an adjacent joint space (Figs. 6-1, 6-4). This creates an injury that prevents full use of the joint and a lame animal regardless of whether the fracture heals or not. Injury to growth plates is a complication of osteomyelitis in the immature patient, with complete or partial cessation of growth leading to shortening or bowing of the affected bone. In cases of paired bones, injury to one bone with cessation of growth in that bone can result in abnormal growth of the other bone. Bone inflammation can be reactivated after a period of quiescence, with a change in the radiographic appearance. This is seen most often in association with fractures that fail to heal normally and continue to have recurrence of drainage tracks. Neoplastic conversion can occur with resulting bone sarcoma. Squamous cell carcinomas have been reported to originate from chronic fistulous tracks in humans.

Axial Skeletal Trauma

7

The axial skeleton includes the mandible, skull, and vertebral column. Often, injuries to the ribs, sternum, and pelvis also are visualized on vertebral radiographs. The *appearance of fractures varies* greatly due to anatomic variations and the nature of the injury. Radiographic techniques may be complicated (Morgan, 1993).

Head Trauma

Radiographs of the head are difficult to evaluate because of breed- and species-dependent differences in shape and size of the head. In an effort to overcome this problem of differences in morphology, the radiographic study ultimately may include closed-mouth lateral, open-mouth lateral, dorsoventral, ventrodorsal, closed-mouth lateral oblique, open-mouth lateral oblique, intraoral ventrodorsal, intraoral dorsoventral, open-mouth ventrodorsal, frontal, basal, and occipital views (Fig. 7-1). For the quality of these views to be acceptable, they have to be performed on an anesthetized patient. In the event of a patient with serious head injury, anesthesia is probably not possible, and therefore only lateral and dorsoventral views of the head can be made to serve as a *survey study*. A more complete radiographic study with additional views might be made at a later time to detect more subtle injuries.

Skull fractures are not easily characterized because of marked changes in the character of the skull. Often, many small, displaced, bony fragments are noted, with the displacement in accordance with the type of injury. If the injury is from a blunt instrument, the bony fragments are depressed and frequently are more difficult to visualize. Bite wounds, due to injury sustained in big dog–little dog altercations, often cause outward displacement of the bony fragments as they are avulsed from their normal location. Because of the

heavy bony tissue in the calvarium of many of the larger, heavier breeds, fractures of these bones are relatively uncommon. This is probably due to the fact that if the injury is severe enough to cause fractures, it probably caused the immediate death of the animal. It also is possible for the trauma to the calvarium to cause permanent or transitory neurologic signs without fracturing the bony structures, and the radiographic examination is negative for fractures.

A common site of trauma in the head of larger dogs is the *frontal region* because of the prominence of these bones in many breeds. Marked displacement of bony fragments of the frontal bone may be noted, and yet the patient presents only with a "headache." The bony fragments usually are comminuted and depressed. These fractures should be considered open fractures because they open into the air-filled frontal sinuses that contain bacteria in large numbers. Another common site of fractures of the head that can be detected radiographically is the *nasal region*. These fractures are evident as lucent lines within the maxilla and premaxilla, but may be difficult to identify because of the associated hemorrhage within the nasal passages that tends to obscure the radiolucent fracture lines (see Fig. 7-1). It is possible to plate fractures of the maxillary and nasal bones. The light cortical bone in the upper jaw heals with a minimal amount of callus, so that fracture healing is more difficult to evaluate clinically. Good occlusion of the teeth can be achieved by surgical reduction of these fractures.

Fractures of the *zygomatic arch* are common because of their lateral prominence on the head, are relatively easy to identify on the ventrodorsal or dorsoventral view, but rarely require treatment. Fractures of the *base of the skull* are difficult to assess and are not treated because of inaccessibility and the severity of the injury.

53

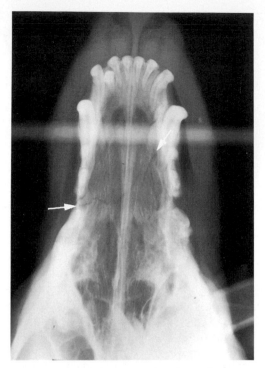

FIGURE 7-1. An open-mouth radiograph showing a maxillary fracture (arrows) in a mature dog. Note retention of an incisor tooth.

Mandibular Fractures

Fractures of the mandible most commonly involve the *body* and are readily identified radiographically because they more closely resemble the fractures of a long bone, with disruption of the heavy cortical shadow. Fractures of the *ramus* of the mandible are much more difficult to identify because the fracture often is obscured by superimposed bony structures. *Symphyseal fractures* of the mandible may be more easily detected by physical examination than by radiographic studies, especially immediately after the injury, because the patient is often in shock and breathing is compromised by hemorrhage within the nasal passages or in the mouth. If possible, an open-mouth radiograph or intraoral placement of a nonscreen film permits more complete evaluation of symphyseal fractures. Treatment of mandibular fractures may require the use of wires, pins, or screws, placed internally. External fixation devices may be used with polymethylmethacrylate instead of wire connecting the bars. It is possible to reduce the fractures using wires placed externally around teeth. Intraoral plastic molds can be fitted to achieve reduction and are held with wires. Because of the shape of the mandible, it is necessary to contour carefully any plate to be used. Screws holding the plate should avoid the roots of the teeth and the mandibular canal (Fig. 7-2). Fractures involving the *temporomandibular joints* need reduction if possible because of their importance in mastication. Pharyngostomy tubes frequently are positioned so that the reduction can be evaluated with the mouth closed and good occlusion of the teeth obtained.

Evaluation of healing is difficult in these lesions because of limitations placed on film positioning as a result of the fixation devices. The mandible has stronger cortical bone that heals in a manner more like that of a long tubular bone, and radiographs made during the healing stage demonstrate an external callus formation. Still, because the mandible is not a weight-bearing bone, the callus formed is much less extensive than that noted in weight-bearing bones.

Dental Fractures

Teeth and periodontal tissues should be assessed radiographically for possible injury after any traumatic injury to the head. These radiographic studies usually are made using oblique views with the film against the outside of the head and with the patient's mouth opened. Nonscreen dental films can be placed within the mouth and an intraoral radiographic technique used to improve the quality of the film; however, placement of the film is more demanding. Even after radiographic detection of the fracture, it may not be possible easily to ascertain the clinical significance of the dental or periodontal injury, or the treatment required.

Spinal Trauma

Trauma to the axial skeleton is a challenge and requires emergency management of the patient. Morbidity and mortality in these patients remain high. It is important for the clinician to advise an owner properly regarding the method of handling the injured patient during the time of transportation to the clinic. The animal should be placed on a stretcher and secured in a fixed position if spinal injury is suspected. If possible, lateral survey radiographs should be made with the animal on this stretcher, to avoid additional movement. Radiog-

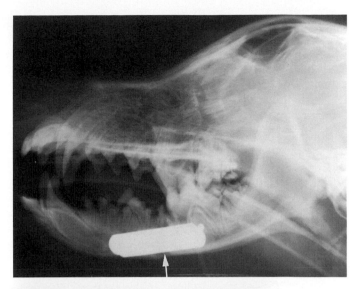

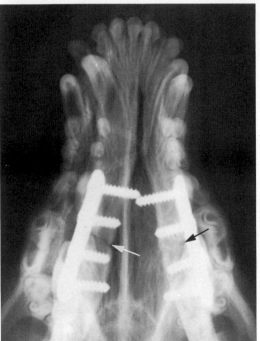

FIGURE 7-2. Lateral and dorsoventral radiographs of a dog with bilateral mandibular fractures (arrows) that have been treated by application of bone plates.

raphy is by far the most useful diagnostic aid in determining the nature and exact location of traumatic injuries of the spinal column (Bailey and Morgan, 1982). Because multiple spinal injuries may be present, the radiographic study should include the entire vertebral column, even if neurologic signs appear to be rather specific in locating the cause of a transverse myelopathy.

Radiology of the spine should provide a method to evaluate traumatic injury to both the vertebral column, including the vertebral segments and the interposed intervertebral discs, and the enclosed nervous structures, which include the: (1) spinal nerves, (2) dorsal roots and dorsal root ganglion of the spinal nerves, (3) ventral roots of the spinal nerves, (4) meninges, and (5) the spinal cord. In

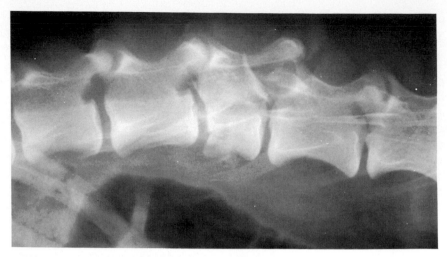

FIGURE 7-3. A lateral radiograph of the lumbar spine showing a badly fractured body of L3 with marked overriding of the fragments.

some cases, the radiograph clearly demonstrates injury to the bone; however, the degree of injury to the nervous structures within the spinal canal is more difficult to ascertain using only conventional, noncontrast radiographic techniques (Fig. 7-3).

It is important to note whether the *location of the radiographic change* coincides with the location of the clinical findings. If the location fails to match the pattern of clinical signs, either the radiographic change is of less clinical significance and not the major cause of the neurologic signs, or

the original injury has enlarged to produce a disseminated lesion such as a myelitis or hematomyelia.

The radiographic changes of a spinal fracture–luxation include: (1) change in width of the disc space, (2) malalignment of the vertebral segments, (3) change in bone density, (4) shortening of the vertebral segments, and (5) end plates that are not parallel (Table 7-1). Fracture lines are difficult to identify because of the minimal amount of cortical bone and because the vertebral fragments may be impacted. Fractures involving the more

TABLE 7-1. Radiographic Changes that May Be Associated with Acute Trauma (Fracture or Luxation) to the Vertebral Column

1. Vertebral luxation indicated by
    a. Scoliosis, kyphosis, or lordosis
    b. Rotational deformity indicated by malpositioned spinal processes
    c. Interruption of the line(s) identifying the floor or roof of the spinal canal
    d. Lateral displacement of vertebral segments
    e. Change in height or width of spinal canal
    f. Change in disc space width
    g. Vertebral end plates not parallel
2. Vertebral fracture indicated by
    a. Malshapen or shortened vertebral segments
    b. Radiolucent fracture lines
    c. Folding cortices of vertebral bodies
    d. Change in size or shape of the vertebral arch
    e. Interrupted end plates
    f. Metallic bullet fragments

dense end plates are more easily recognized than those involving trabecular bone (Bailey and Morgan, 1982).

Narrowing of the disc space may follow a nuclear prolapse or disc herniation due to acute trauma or a vertebral luxation. Narrowing of the disc space may not be uniform, but instead may be in the form of a wedge. A *disc space may widen* as a result of luxation, and malalignment of the adjacent vertebral bodies may be present as well.

Malalignment of intact or fractured vertebral segments is one of the easiest radiographic changes to identify after trauma; a line is drawn along the floor or roof of the spinal canal or the ventral aspect of the vertebral bodies, and any offset suggests malalignment. Malalignment can be directed laterally, creating a scoliosis, in which case the radiographic evaluation must be made on a ventrodorsal or dorsoventral radiograph. On this view, a line is drawn along the inner aspect of the adjacent lamina. Instability between adjacent segments in either plane can be accentuated by the careful use of stress radiography. Another error in alignment may result in a rotational deformity, and is evident in fracture–dislocation cases. The malalignment of the tips of the spinous processes can be seen on either the ventrodorsal or dorsoventral views. Trauma can cause malalignment with the diagnosis made easier if there is associated luxation at the disc space with a change in disc width, or if end plates of adjacent vertebral segments are not parallel.

Increased bone density without associated change in shape of the bone takes place with impaction of the trabeculae. A *decrease in bone density* may result from fractures that cause separation of the bony fragments.

Shortened vertebral segments are seen due to impaction of the bone secondary to the trauma. The compression fractures also may change the size and shape of the vertebral body as well as increase its density due to the bony tissue collapse.

Adjacent *end plates* are normally parallel, but with luxation they assume a position in which they form a wedge shape.

Rib and Sternal Trauma

Fractures of the ribs occur commonly as the result of automobile injury, falls, crushing bite wounds, or sharp blows to the chest wall, and are detected by noting displacement of the fracture fragments on the radiograph. If fragment ends overlap, a more dense shadow is noted, whereas if fragment ends are separated, a radiolucent space is noted between the fragment ends. In both situations, a definite alteration in the normal contour of the affected rib may be present. Depending on the nature of the trauma, the fractures may be similarly placed on both sides of the chest wall, suggesting penetrating bite wounds. With more blunt trauma, the fractures involve adjacent ribs in only one chest wall. If there are two fractures each in adjacent ribs, a segment of thoracic wall is separated and moves in a paradoxical direction with respiration. This is called a *"flail" chest* and seriously compromises the patient's ability to breathe. In most trauma patients, the soft tissue injury to the chest wall and lung is more important than the rib fractures.

Sternal fracture–dislocations are unique injuries. The sternal segments are not weight bearing, and clinical signs associated with their injury often are minimal. Because of location, injury to the sternal segments indicates possible associated injury to the chest wall, and the detection of injury to the sternum mandates careful attention to the remainder of the thorax and examination for more clinically significant lesions.

Appendicular Skeletal Trauma

8

Diagnostic radiology is conveniently used in clinical practice for the diagnosis of fractures of long bones, evaluation of reduction, and determination of the prognosis of fracture healing. A fracture within a long bone is most easily defined as a lesion causing an interruption of the continuity of the bone resulting from stress that is beyond the capacity of the bone to withstand. The radiographic study should include the joints both proximal and distal to the injury, and should include two orthogonal views. Only with a complete radiographic study can the full character of the fracture be determined and the possible involvement of the adjacent joints be evaluated. Both the character of the soft tissue injury as well as the nature of the bony injury should be evaluated on the radiographs (Morgan, 1993).

Scapular Fractures

Fractures of the proximal portion, the spine, or blade of the scapula often are caused by a car passing over the patient's body, and result in linear fracture lines with bent or folded fragments. Often the bone may split like a wooden shingle with separation of the spine. Diagnosis of these fractures is difficult radiographically because there are no strong cortical shadows that are interrupted. Rotation of the body in making the dorsoventral or ventrodorsal view helps to project the blade "on end" and makes fragment displacement easier to see. The lateral view is confusing because the bone is not dense enough for detection of linear fracture lines. The lateral view is better for evaluation of fractures of the scapular neck. If the fragments overlap, the increased density can be identified on either view. Because of the flattened character of the bone, most fractures result in superimposition of bony fragments, with the appearance of lines of increased density within the bone instead of the expected radiolucency.

The clinical significance of some of these fractures is minimal unless the neck is involved. However, fractures involving the blade frequently are quite painful and may require treatment to relieve the pain, because the patient is not bearing weight on the affected thoracic limb. The presence of an articular component through the glenoid cavity increases the clinical significance considerably. Because of this, the glenoid cavity should be examined carefully for intra-articular fracture lines. Without treatment, the patient faces the possibility of secondary joint disease. If the fracture involves only the blade or spine, or both, the resulting malunion may affect locomotion adversely in the athletic patient. The *surgical approach* is relatively easy, with the incision made over the spine with displacement of the supraspinatus muscles. This provides an opening to apply a plate or wires. Fractures of the scapula may occur with extreme abduction of the limb and possible injury to the brachial plexus, making the neurologic examination important.

A specific type of injury in the dog is an *avulsion of the supraglenoid tubercle* with displacement after the fracture and the appearance of a free fragment within the soft tissues cranially on the radiograph. This is commonly seen in the younger patient. Separation seems to follow the patient running into some object and striking the shoulder, or it may occur when the dog is running hard and drives off of the affected limb. The fracture may require wires to achieve reduction and stabilization using a lag screw.

Humeral Fractures

Radiography of the forelimb can be performed using a horizontal beam with the patient lying in lateral recumbency for both radiographic views (Fig. 8-1). The beam can be directed craniocaudal or caudocaudal. This technique avoids having to

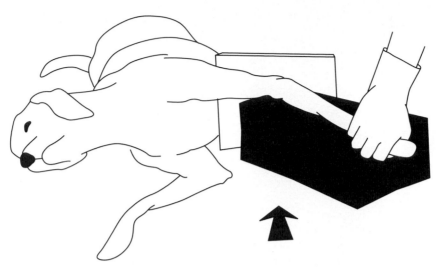

FIGURE 8-1. Drawing of a dog positioned in lateral recumbency for a craniocaudal view of the upper forelimb using a horizontal x-ray beam. The limb can be extended to obtain a more accurate view of the lower forelimb, if necessary, without great pain to the patient.

position the fractured bone with the patient in sternal recumbency for a craniocaudal view. This usually is painful because the limb needs to be extended and the patient is bearing partial weight on the limb.

Most fractures of this bone involve the midshaft and distal portion. The midshaft fractures typically are spiral in character, often with comminution or butterfly fragments, whereas the distal fractures involve the distal condylar region creating, "T" or "Y" fractures that are articular. Physeal fractures of the proximal and distal humerus are either type I or II. The possibility of radial nerve injury needs to be considered. Oblique radiographic positioning may assist in determining the exact nature of these fracture lines. In the skeletally immature patient, ossification centers for the capitulum and the trochlea may separate with trauma, forming two fragments that need to be reattached. The reunited epiphysis also will need to be fixed to the metaphysis. In the mature patient, the fracture line often follows the original cartilaginous site of separation between the two ossification centers. These fragments are reduced through use of a transcondylar screw. There is ample bone to place this screw in the dog; in the cat, however, little space is available for screw placement in reduction of the condylar fragments. Exact reduction of the condylar fragments is a re-

quirement, with congruence necessary to avoid secondary joint disease.

Plating often is used for fracture repair, with the plate placed on the lateral, medial, or cranial surface of the bone, depending on the fracture type and number of fracture fragments. Care must be taken to avoid injury to the radial nerve and the brachialis muscle. To gain exposure to the distal humerus, a triceps tenotomy often can be used instead of an olecranon osteotomy. A variety of other fixation techniques is available to the surgeon.

Radial Fractures

This is one of the more commonly fractured bones in the dog and cat, and fractures of almost any description can occur. The majority of radial fractures also have an associated ulnar fracture. Most fractures occur in the midshaft and distally, and rarely include an articular component. Both craniocaudal and lateral radiographic views are necessary to determine accurately fragment description and displacement. Both the dog and cat protect the injured limb rather well by "tripoding" the body after the injury so that additional soft tissue injury is not common. Physeal fractures may occur either proximally or distally and are not always detected at the time of injury, but are detected

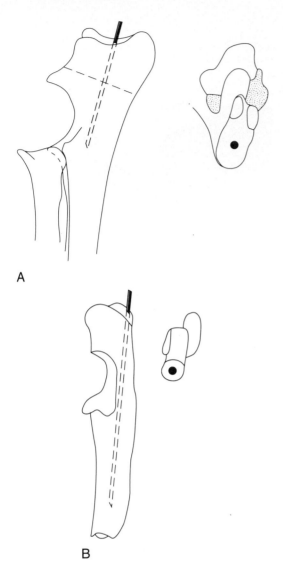

A

B

FIGURE 8-2. Drawings of the proximal ulna in the dog (A) and the cat (B). The location of the entrance point of the pin is located in each drawing, as well as the site of the pin after placement. Note the larger bone in the dog with a marked curvature toward the lateral side. The bone is smaller in the cat and is much straighter, and the medullary cavity is much smaller.

rather at the discovery of delayed growth or closure of a growth plate.

Small and, particularly, toy breeds are liable to difficulty in healing of distal radial fractures, with a *secondary disuse osteopenia* resulting. This may be followed by actual bone atrophy with severe penciling of the bone ends. Usually this is the result of failure to obtain a degree of stabilization that permits the patient to use the limb partially, thus avoiding atrophy. Treatment with a small plate and cancellous graft is extremely important in this clinical situation.

Ulnar Fractures

Ulnar fractures are common and frequently occur in conjunction with radial fractures. Repair of the radial fractures usually produces sufficient stability, and the concurrent ulnar fractures are left untreated. A separate group of fractures involves the olecranon or the olecranon process in the immature patient. The pull of the triceps muscle creates distraction of the fracture fragment. These fractures often may be articular with all of the inherent problems associated with this type of injury. The proximal end of the ulna has a markedly different appearance in the dog and cat (Fig. 8-2). In the dog, the proximal ulna is rather large, with opportunity for placement of an IM pin. Curvature of the proximal ulna is to the lateral side, so a pin directed from the proximal end of the olecranon easily passes through the medial cortex. In the cat, the proximal ulna is smaller, but it is much easier to place an IM pin within the medullary cavity because of the straightness of the bone. The problem in pin placement in the cat comes with small size of the medullary cavity and the possibility that the pin may pass into the trochlear notch. Fractures of the styloid process may be seen in combination with a radial fracture, and are the cause of antebracheocarpal joint instability.

The conformation of the *distal ulnar growth plate* is remarkably different from the other growth plates in the dog. Instead of a transverse plate, the shape is that of an inverted cone in which shearing forces are converted to a crushing-type injury with compressive forces that cause severe injury. This means that in the absence of posttraumatic radiographic changes, the distal ulnar growth plate still can be injured to the extent that delayed growth or early physeal closure may occur.

Carpal, Metacarpal, and Phalangeal Fractures

The small carpal bones often are crushed, and radiography should include four views to evaluate completely for the presence and nature of these fractures. Stress views, especially hyperextension, may help in separating overlying bones so that fractures and sites of instability can be seen more easily. Because of variations in morphology and the absence of prominent cortical shadows, carpal fractures are difficult to identify. Injury to the accessory carpal bone may be associated with a ventral avulsion fracture, a dorsal avulsion fracture due to the attachment of the ulnocarpal ligament, a distal fracture at the attachment of the carpal–metacarpal ligament, or an avulsion fracture of the proximal tip of the accessory carpal bone displacing proximally because of the attachment of the flexor carpi ulnaris muscle tendon. The fractures may result in the displacement of the accessory carpal bone proximally or distally, or it may displace proximally due to tearing of the two unnamed distal ligaments that originate from the rounded free end of this bone and ultimately attach to the fourth and fifth metacarpal bones.

Metacarpal bone fractures are common in both the dog and cat. If the fracture involves the second and fifth metacarpal bones, splinting of the foot is sufficient because perfect repair is not required. Fractures of the third and fourth metacarpal bones need better reduction because they are important in ambulation, especially in the athletic dog. It is possible to use cerclage wire as primary fixation in conjunction with the use of an external cast or splint in the smaller patient. Intramedullary pins are used successfully, with the best pinning technique being to introduce the pin on the dorsal surface of the distal metacarpal bone and advance it in a retrograde direction. As a result, the metacarpophalangeal joint can be avoided. It is not possible to use an IM pin in combination with interfragmentary screws because of the small size of the medullary cavity. In some cats and toy breeds, the intramedullary canal is nonexistent. Miniplates are used frequently in treatment of metacarpal fractures.

Proximal and middle phalanges are tubular and fractures resemble fractures of larger tubular bones. The distal phalanx is difficult to position for radiography, and a "paddle" technique forcing the foot flat against the table top may assist in achieving better positioning. Fractures of the proximal sesamoid bones at the metacarpal–phalangeal joints are a distinct type of injury to the foot and are important clinically. They are numbered from one to eight beginning medially on the second digit, and lie within the digital flexor tendon where they can be reached rather easily for complete excision or fragment removal using direct surgical exploration.

Pelvic Fractures

Fractures of the pelvis are present in 20% to 30% of traumatic injuries in the dog and cat. Fractures of the pelvis are unique because of its basic anatomic structure, which leads to the "ring" or "box" concept for understanding the distribution of the pattern of fractures or luxations. Because of this unique morphology, trauma to the pelvis usually results in three separate but related lesions. Often, there are three fractures, with one each involving the ilium, pubis, and ischium on the same side, or involving both sides. Another frequently seen pattern is two bony fractures plus a sacroiliac or pubic symphyseal separation. A less common pattern of injury occurs with separation of the two halves of the pelvis, with fracture of one sacroiliac joint and separation of the pubic symphysis. It also is possible to have an acetabular fracture plus a sacroiliac separation that divides the pelvis into two sections.

In cats, both *sacroiliac joints* often are luxated with separation of the intact bony pelvis from the sacrum. It is important in both dogs and cats with sacroiliac injury to question whether the sacrum is fractured, even though the wings of the sacrum are difficult to evaluate radiographically. Sacral fractures are clinically important because of potential physical instability as well as the possibility of injury to the cauda equina. In the placement of lag screws to stabilize sacroiliac fractures, it is critical to obtain good purchase within the sacrum. A single screw that reaches 60% of the distance across the sacrum is sufficient for most dogs.

Fractures often occur in conjunction with an *unstable hip joint*. Fractures may enter the acetabulum and destroy the contour of that joint. Implosion-type fractures cause marked acetabular deformity that requires treatment if good joint function is to be preserved and a future dystocia in the intact female and obstipation in both sexes are to be avoided. Even with fragment healing, secondary joint disease may occur because of the nature of the disruption to the contour of the joint surface. Reduction of fractures involving the roof of the acetabulum is important, whereas fractures of the floor of the acetabulum need not be so carefully reduced because they do not affect the weight-bearing surface. In a study of dogs with caudal one-third acetabular fractures managed with cage rest, healing and subsequent hip function were found to show unsatisfactory results, and

TABLE 8-1. Radiographic Features of Pelvic Trauma
(Occurring Singly or in Combinations)

1. Pattern of fractures involving the ileum, pubis, and ischium on the same side or involving both sides
2. Bilateral or unilateral sacroiliac separation
3. Pubic symphyseal fractures
4. Acetabular fractures
5. Avulsion of ilial and ischial crest apophyseal centers
6. Sacral fractures
7. Sacrococcygeal or coccygeal luxation
8. Intrapelvic hemorrhage
9. Cranial displacement of the hemipelvis
10. Narrowed pelvic canal
11. Proximal femur fractures
12. Coxofemoral luxation

thus even caudal acetabular fractures should be managed surgically if possible (Boudrieau and Kleine, 1988). The radiographic features of pelvic trauma are listed in Table 8-1.

Additional injury to the neighboring bony structures is possible with pelvic fractures. This includes avulsion of the apophyseal centers of the ilial and the ischial crest and is found in the younger patients in whom skeletal maturation has not occurred, and usually is without unique clinical signs. Because of the closeness of the proximal femur, any patient with an injury to the pelvis needs to have the femoral head, femoral neck, and greater trochanter closely evaluated by physical and radiographic examination. Traumatic separa-

tion of the sacral segments or fracture–luxation of the caudal vertebral segments often occurs with trauma of the pelvis and needs to be detected, because this injury often is associated with paresis or paralysis of the tail.

Femoral Fractures

The femur is one of the most frequently fractured bones, causing 20% to 25% of all fractures, encompassing a wide variety of fracture types. In the young patient, type I physeal fractures involve the femoral head. In the immature patient, the slipped capital epiphysis usually loses its blood supply and undergoes necrosis. In the adult patient, it is im-

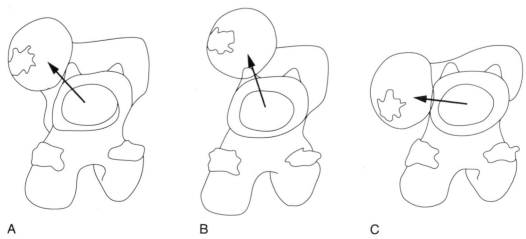

A   B   C

FIGURE 8-3. Drawing of the femur as seen from the proximal end with the head and neck in a normoversion position (A), an anteversion position (B), and a retroversion position (C). The angle of the head and neck relative to the shaft is identified (arrows).

portant to note whether femoral neck fractures are intracapsular or extracapsular, because this determines something of the possibility of disruption of the vascular supply to the femoral head and neck, leading to aseptic necrosis.

Midshaft fractures result in rotation of the proximal femur into a position of anteversion or retroversion of the femoral head and neck. This needs to be evaluated on postoperative radiographs because this type of malpositioning can be significant clinically (Figs. 8-3, 8-4). With anteversion, the limb is forced into internal rotation when walking or running, whereas with retroversion, the limb is forced into external rotation. Fractures that heal in malposition often can result

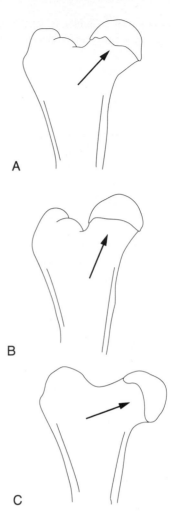

A

B

C

FIGURE 8-5. Drawing of the femur as seen craniocaudally showing the femoral head and neck in a normal position (A), in a position with valgus deformity (B), and in a position with varus deformity (C). The angles are identified (arrows). The relationship of the femoral neck to femoral shaft determines how the femoral head can fit into the acetabulum.

FIGURE 8-4. Drawing of the femur after fracture repair as seen on the lateral radiographic view. A line drawn along the cranial cortical shadow bisects the femoral head in the average dog or cat when the femoral head and neck are in a normal position (line). Thus, a normal anatomic position is slightly anteverted.

in the femoral head and neck being in a position of varus or valgus deformity (Fig. 8-5).

Distal physeal fractures are often type II. Articular fractures are uncommon proximally but do occur distally with separation of a femoral condyle. Midshaft fractures result in marked displacement and overriding of comminuted or butterfly fragments.

Positioning for radiography of the femur is difficult because of problems in extending the fractured limb when the patient is in dorsal recumbency. It is possible to make this view with the limb in a flexed position, referred to as a "frog-leg" position. It also is possible to obtain a craniocaudal view by putting the patient in a "sitting" position and pulling the foot distally. Use of a horizontal beam is perhaps the easiest method for obtaining the craniocaudal or caudocranial view, but may not be possible with some x-ray machines. Another radiographic technique is to diagnose the character of the fracture partially on the lateral view and delay making the second orthogonal view until later, when the patient is anesthetized for surgery. Radiography of the opposite limb often is used to determine the correct length to be achieved in stabilization of the fractured bone. Also, the degree of distal bowing of the femur can be determined in estimating the length of the IM pin or plate required.

In using a plate for proximal femoral shaft reduction, one of the most proximal screws should be at an oblique angle entering the femoral neck, and not transverse relative to the femoral shaft. Intramedullary pins are introduced into the trochanteric fossa in a normograde direction and are driven distally to cross the fracture site. Morphologically, there is a great difference between the femur in the cat and dog, with the femur in the dog being a curved bone whereas in the cat it is much straighter. This means that the use of proportionately larger IM pins in the cat is indicated. In the dog, the surgeon must focus attention to the smaller diameter of the medullary cavity. In addition, to seat the distal tip of an IM pin in the dog, the distal fragment needs to be overcorrected beyond anatomic position, with cranial angulation so that the pin seats distally (Fig. 8-6).

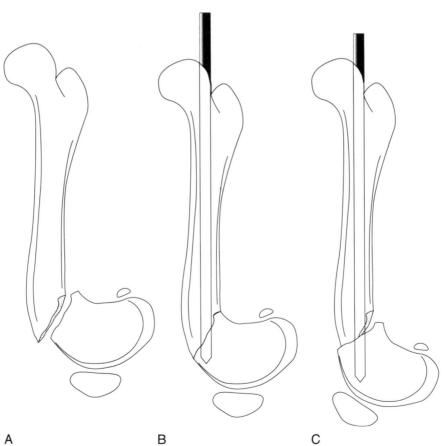

A   B   C

FIGURE 8-6. Drawing of the femur as seen on a lateral view showing a distal femoral fracture. The displaced fragment is shown (A). Repositioning the fragment in an anatomic position results in placement of the IM pin with only the tip entering the distal fragment (B), whereas overreduction of the fragment permits the IM pin to enter the fragment more deeply (C).

Tibial Fractures

Fractures of the tibia are common and comprise many different types. In the immature patient, type II physeal fractures of the proximal epiphysis are most common. Replacement of avulsion fractures of the tibial tubercle is particularly important to ensure correct positioning of the patella and normal use of the patellar ligament. Midshaft fractures often are oblique and comminuted or have large butterfly fragments. Often there is fragmentation at the fracture site, and the use of a plate in a buttress mode is strongly recommended. In determining the correct length of the bone, it is helpful to radiograph the opposite limb and measure the tibia to achieve the sought-after length. Distally, fractures of the medial malleolus may be associated with tibiotarsal luxations and destruction of the ankle mortise.

Fibular fractures often are found in association with tibial fractures. Unfortunately, if the fibula is intact in the dog, it does not provide enough splinting to be of great value in stabilization of the tibial fracture. The fibula is much stronger in the cat and has value in splinting tibial fractures. Fractures of the proximal fibula need reducing because of the attachment of the lateral collateral ligament to the proximal tip.

Tarsal, Metatarsal, and Phalangeal Fractures

Diagnosis and fractures are similar to those noted in the distal portion of the forelimb. A tension-band apparatus is used successfully to treat calcaneal fractures. Exposure of the tarsus can be accomplished by performing an osteotomy of the distal fibula with subsequent replacement.

Joint Injury

9

Intra-articular Fractures

Fractures that enter the joint space result in disruption of the articular cartilage and may result in healing that leaves a roughened or irregular articular surface. This irregularity of the cartilaginous surface causes joint instability with development of a secondary arthrosis. The degree of original injury to the joint surface plus the level of anatomic restoration determines the degree of arthrosis that develops. Features such as the size and weight of the patient and the degree of daily or weekly exercise also determine the progression of the resulting arthrosis. Articular injury to the immature patient has a much greater chance of healing with a good clinical result because of the high rate of turnover of bony tissue in young patients. It is important that the clinician understand this concept because it makes it possible to forewarn the owner that a fracture–luxation that is otherwise healing may have secondary changes with long-lasting clinical significance. Radiology of intra-articular fractures may be difficult (Payne, 1993).

The first and most important principle of intra-articular fracture repair is that reduction must be anatomic. Another principle is the need for absolute rigid stabilization of the fragment(s). A final principle is that a range of motion and weight-bearing capability must be obtained as soon as possible after surgery to provide for optimal cartilage healing as well as optimal healing of the periarticular soft tissues (Payne, 1993).

Fracture–Luxation

Fracture–luxation is a combination of injury to the bone as well as injury to the joint. The fracture may be articular or extra-articular. The clinical significance of an articular fracture is greater if reduction of the luxation and repositioning of intra-articular bony fragments does not result in a normal anatomic relationship. Any resulting instability or malalignment of the joint surfaces leads to the development of posttraumatic arthrosis. Injury of this type is more important clinically in a heavy, athletic patient if the injury involves an important weight-bearing joint. If the fracture is extra-articular, the avulsion of the tendinous and ligamentous attachments may not be appreciated and the lesion undertreated. If the fracture is intra-articular, it is possible for a bony fragment to be without a blood supply.

Hemarthrosis

The immediate results of the crushing of the synovial membrane of a joint or its communicating bursa are hemorrhage and edema. The amount of fluid produced is large, and out of proportion to the injury, but this transudate usually is absorbed within a few days. However, if the injury is more violent, blood may collect within the joint space, interrupting the nutrition of the articular cartilage. Hemarthrosis is difficult to recognize radiographically because the distention of the joint capsule may not be distinguished from intra-articular effusion, or from the surrounding extracapsular swelling and edema.

Intra-articular Gas

Curvilinear or streaky radiolucent shadows within a joint are due to the presence of gas, usually air. In the patient with acute injury, the shadows may follow an open injury in which air gains direct entrance into the joint. Although uncommon, this would be an important radiographic finding indicating injury to the joint capsule or surrounding bone.

Posttraumatic Septic Arthritis

Septic arthritis may follow trauma that resulted in a tearing of the joint capsule associated with an open injury, or may follow an operative procedure that involved the opening of a joint. The first radiographic changes are nonspecific and are those of soft tissue swelling with capsular distention. Later, the subchondral bone becomes osteopenic and a fluffy pattern of periosteal new bone is seen equally on both sides of the joint. The new bone production may be lost within the early callus formation around the fracture site. The infection usually is detected clinically before the radiographic signs are conclusive (Figs. 9-1, 9-2).

Luxation, Subluxation, and Dislocation

Instability of a joint is referred to as luxation, subluxation, or dislocation, depending on the degree of instability and separation of the articular surfaces. Detection of injury to the joints on physical examination may be relatively easy in most trauma patients. Unfortunately, positioning of the limb for radiographic study may result in reduction of a luxation or instability and cause the joint to appear more nearly normal. *Stress studies* are therefore necessary to fully appreciate radiographically the limits of joint luxation or instability. This is a common injury involving the palmar carpal ligaments and fibrocartilage (Gambardella and Griffiths, 1982). It is important to understand that strain or sprain of the soft tissues surrounding a joint often requires a much longer time to heal than does a fracture, with most injury to bone tissue. The manner of healing of the joint injury is greatly affected by the degree of success in reducing the luxation and maintaining normal anatomic alignment of the bones during soft tissue repair.

Shoulder Joint

The shoulder is unique anatomically because there are no definite ligaments that maintain the humeral head and glenoid cavity in persistent position. Consequently, on radiographs made with abduction or traction, the joint space appears to increase in width, suggesting luxation. Radiography of the contralateral limb may be helpful in determining the normal range of motion. In some patients, it may be difficult to determine whether a chronic luxation is posttraumatic or congenital.

Elbow Joint

The elbow joint is relatively easy to judge radiographically because it is a tightly fitting joint and luxation results in marked separation of the ends of the bones, with or without associated fractures. Small avulsions or chip fractures often are associated with elbow joint luxation.

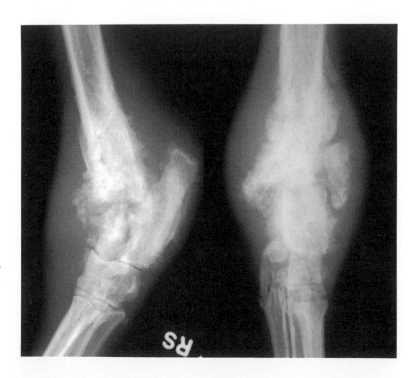

FIGURE 9-1. Radiograph of the foot of an 8-year-old female Irish Setter with an injury to the lower limb. An open track developed and the lesion was curetted. The malleolar fracture became relatively unimportant with the development of infectious arthritis. The tibiotarsal joint is destroyed.

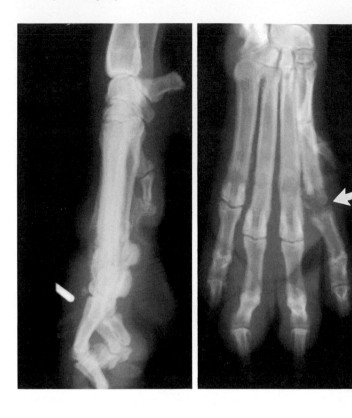

FIGURE 9-2. Radiographs of the foot of a mature shepherd mixed breed with destruction of the carpometacarpal joint of the second digit (arrow). Infectious arthritis followed having the foot caught in a trap overnight.

Antebracheocarpal, Intercarpal, Carpometacarpal, Metacarpophalangeal, and Interphalangeal Joints

Injury to these joints is best evaluated on hyperextended views or other stress views that demonstrate the degree of ligamentous injury, associated fractures, and the resulting joint laxity (Morgan, 1993). In patients with injury to the collateral ligaments, stress views can be made with the foot forced into a valgus or varus position. Interphalangeal joints are more difficult to evaluate radiographically and it may be necessary to force the foot into contact with the cassette using a plastic or wooden paddle.

Hip Joint

The hip joint is relatively easy to evaluate posttraumatically because the femoral head tends to separate completely from the acetabulum and luxates to a position dorsocranially along the ilium. It is important to examine the radiograph for associated acetabular fractures. A difficulty in diagnosis occurs in the patient in which the acute trauma is superimposed over a dysplastic hip joint that was

previously unstable due to joint capsule stretching and flattening of the acetabular cup. The same problem occurs with a luxation detected in a patient with chronic changes secondary to aseptic necrosis of the femoral head. If the patient is in pain during the radiographic procedure, it is possible to make the study with the legs in a fully flexed position (frog-legged position).

Stifle Joint

The stifle joint is more difficult to evaluate radiographically compared to joints that have articular surfaces that are more congruent, such as the elbow or hip joint. In tightly fitting joints, the components are either "in or out," and diagnosis of injury is easier. Because the joint space between the "round" femoral condyles and the "flattened" tibial plateau normally changes in width, it is difficult to visualize what is the normal radiographic appearance. These bones "fit" only because of the radiographically invisible menisci. Therefore, luxation of the stifle joint is more difficult to ascertain radiographically, although it is possible to attempt to generate cranial displacement of the tibia or separation of the bones using stress techniques (drawer movement). Craniocaudal stress demonstrates cruciate ligament injury, whereas mediolateral stress demonstrates collateral ligament injury.

Tibiotarsal, Intertarsal, Tarsometatarsal, Metatarsophalangeal, and Interphalangeal Joints

The tibiotarsal joint is a tightly fitting joint, and instability usually results from fracture of one of the malleoli with resulting luxation that is clearly evident radiographically. Injury to this joint may be seen more clearly radiographically by stressing the foot into valgus or varus positions. Intertarsal and tarsometatarsal joints usually require stress radiographs to evaluate fully the degree of joint instability and associated fractures. These joints are best evaluated in hyperflexed and hyperextended positions. If the injury is extensive enough, the bones remain permanently malpositioned.

Pathologic Fractures

10

Introduction

Pathologic fractures are uncommon in animals, may involve a single bone or multiple bones, and may be due to malignant disease, or, more commonly, to nutritional disease. These fractures often are not complete, but instead show a "folding" or "bending" pattern.

Spinal Pathologic Fractures

In the spine, fractures of the vertebral bodies are of a compression type with the ventral cortex "folded." Radiographic diagnosis usually is made by noting shortened vertebral bodies with a thinner vertebral end plate and preserved disc spaces. The shortening of the vertebral bodies often is associated with a kyphosis or scoliosis.

Multiple fractures often are associated with secondary nutritional hyperparathyroidism in a younger patient on an unbalanced diet. Bone density in these rapidly growing animals is minimal even under the best nutritional conditions, and the determination of what is pathologic is difficult. One test that can be used is to examine the radiograph, looking for the ventral cortex of the vertebral bodies. Normally, these rather sharply defined, dense lines represent the cortex and contrast with the less dense cancellous bone in the vertebral body and the less dense perivertebral tissues. If this line is missing, it is an indicator of osteopenia.

Mandibular Pathologic Fractures

Pathologic fractures of the mandible usually are associated with renal osteodystrophy or chronic periodontal disease and may be remarkable for the degree of osteopenia present. Similar loss of bone density is difficult to identify within the maxilla and premaxilla, and fractures are less common.

Long Bone Pathologic Fractures

Pathologic fractures in the long bones may be multiple and due to nutritional problems, or found in one bone in association with a solitary bone lesion. They are most easily detected by the abnormal conformation of the bones. With systemic disease, bent or folded cortices cause an increase in bone density due to the "doubling" of the cortical bone. Careful examination of all of the bones may show, in addition, older healed fractures. The bent cortical shadows present in the acute fractures repair with an uneven thickness. In addition, there is a generalized thickness of the cortex on the concave side because it is strengthened from greater weight bearing.

With a solitary lesion, cortical destruction usually is evident along with a soft tissue mass. The pathologic fracture often results in compaction of the fragments, in contrast with the clean fracture line found in healthy bone.

Soft Tissue Trauma

11

Radiographs rarely play a role in evaluation of isolated soft tissue injuries unless there is a question of retained foreign matter. Skin lacerations may cause the accumulation of gas within the soft tissues, making radiographic evaluation more difficult. When there is removal of large portions of skin and underlying muscle, tendon, and ligaments, as is seen in abrasive injuries, the radiographic evaluation must take into account the highly significant soft tissue injury.

Soft Tissue Swelling

Soft tissue swelling is difficult to detect on radiographs of the axial skeleton or on proximal portions of the limbs, and is more easily seen distally around the feet. When detected, soft tissue swelling is a good indicator of the location of the injury. A good rule is "to permit the soft tissue swelling to direct your attention to the lesion in the underlying bone." If a bone lesion is present, it is often in this location. Swollen joints often are associated with a strain or sprain in the absence of joint instability or fracture.

Soft Tissue Gas

Gas within the soft tissues is common in trauma cases in which the integrity of the skin has been broken, with the air accumulating in subcutaneous tissues or following the muscle bellies. Often it is not convenient to examine physically the fractured limb and identification of the gas on the radiograph may be the first indication of the open nature of the fracture. A most common cause of soft tissue gas noted radiographically is iatrogenic (postsurgical) (Fig. 11-1).

Soft Tissue Foreign Bodies

Any radiopacity within the soft tissues is suggestive of a penetrating foreign body, and has significance. Careful examination also should be made for an additional underlying bone injury. Areas of calcification within the soft tissues are indicative of chronic injury and are not associated with an acute injury. If localization is required, it is possible to place radiopaque skin markers and make additional radiographic exposures. It also is possible to insert radiopaque needles to serve as locators of the foreign body.

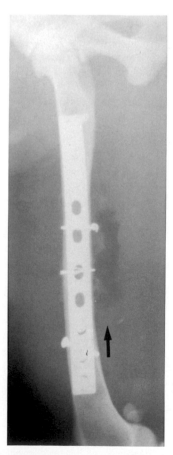

FIGURE 11-1. Postoperative lateral radiograph of the femur of a mature cat showing the appearance of air (arrow) within soft tissues as a result of the surgical procedure.

References

Backgren AW, Olsson S-E. Intra-articular fractures following traumatic coxofemoral luxation in the dog and cat: A radiographic and pathological study. Nord Vet Med 13:197–204, 1961.

Bagnall BG. Reaction of dissimilar metals used in orthopaedic surgery: A report of two cases. J Small Anim Pract 13:201–206, 1972.

Bailey CS, Morgan JP. Disease of the spinal cord. In: Ettinger SJ (ed): *Textbook of Veterinary Internal Medicine*, 2nd ed. Philadelphia: WB Saunders, 1982:532–607.

Behrens F. A primer of fixation devices and configurations. Clin Orthop 241:5–14, 1989.

Berger PE, Ofstein RA, Jackson DW, Morrison DS, Silvino N, Amador R. MRI demonstration of radiographically occult fractures: What have we been missing? Radiographics 9:407–436, 1989.

Bick EM. Structural patterns of callus in fractures of the long bones. J Bone Joint Surg [Am] 30:141–150, 1948.

Birchard SJ, Bright RM. The tension band wire for fracture repair in the dog. Compend Cont Ed 3:37–41, 1981.

Björck G. Transfixationsgipsning vid behandling av frakturer hos mindre husdjur. Nord Vet Med 4:89–115, 1952.

Boudrieau RJ, Kleine LJ. Non-surgically managed caudal acetabular fractures in the dog: 15 cases (1979–1984). J Am Vet Med Assoc 193:701–705, 1988.

Bowerman JW, Hughes JL. Radiology of bone grafts. Radiol Clin North Am 13:67–77, 1975.

Braden TD, Brinker WO. Radiologic and gross anatomic evaluation of bone healing in the dog. J Am Vet Med Assoc 169:1318–1323, 1976.

Brearley MJ, Houlton JEF. Galvanic corrosion due to stainless steel implants of differing composition in a dog. J Small Anim Pract 24:489–494, 1983.

Brinker WO, Flow GL, Braden T, Noser G, Merkley D. Removal of bone plates in small animals. J Am Anim Hosp Assoc 11:577–586, 1975.

Brinker WO, Piermattei DL, Flo GL. *Handbook of Small Animal Orthopedics and Fracture Treatment*, 2nd ed. Philadelphia: WB Saunders, 1993.

Butt WP. The radiology of infection. Clin Orthop 96:20–30, 1973.

Caywood DD, Wallace LJ, Braden TD. Osteomyelitis in the dog: A review of 67 cases. J Am Vet Med Assoc 172:943–946, 1978.

Clayton-Jones DG, Vaughan LC. Disturbance in the growth of the radius in dogs. J Small Anim Pract 11:453–468, 1970.

Compere EL. Avascular necrosis of large segmental fracture fragments of the long bones. J Bone Joint Surg [Am] 31:47–54, 1949.

Dallman MJ, Martin RA, Seif BP, Grant JW. Rotational strength of double-pinning techniques in repair of transverse fractures in femurs of dogs. AJVR 51:123–127, 1990.

DeAngelis MP. Causes of delayed union and nonunion of fractures. Vet Clin North Am 5:251–258, 1975.

DeVas MB. Compression stress fractures in man and greyhound. J Bone Joint Surg [Br] 43:540–551, 1961.

Fox SM. Premature closure of distal radial and ulnar physes in the dog. Compend Cont Ed 6:128–138, 212–221, 1984.

Gambardella PC. Full cerclage wires for fixation of long bone fractures. Compend Cont Ed 2:665–671, 1980.

Gambardella PC, Griffiths RC. Treatment of hyperextension injuries of the canine carpus. Compend Cont Ed 4:127–132, 1982.

Gannon JR. Stress fractures in the greyhound. Aust Vet J 48:244–250, 1972.

Gibson KL, vanEe RT, Pechman RD. Femoral capital physeal fractures in dogs: 34 cases (1979–1989). J Am Vet Med Assoc 198:886–890, 1991.

Gillies C. The x-ray diagnosis of fractures. Clear Images 25–34, January 1990.

Göthman L. Vascular reactions in experimental fractures: Microangiographic and radio-isotopic studies. Acta Chir Scand [Suppl]:284, 1961.

Göthman L. Local arterial changes caused by surgical exposure and the application of encircling wires (cerclage) on the rabbit's tibia: A microangiographic study. Acta Chir Scand 123:9–16, 1962a.

Göthman L. Local arterial changes associated with experimental fractures of the rabbit's tibia treated with encircling wire (cerclage): A microangiographic study. Acta Chir Scand 123:17–27, 1962b.

Göthman L. Local arterial changes with diastasis in experimental fractures in the rabbit's tibia treated with intramedullary nailing: A microangiographic study. Acta Chir Scand 123:104–110, 1962c.

Harris NH, Kirkaldy-Willis WH. Primary subacute pyogenic osteomyelitis. J Bone Joint Dis [Br] 47:526–532, 1965.

Hinko PJ, Rhinelander FW. Effective use of cerclage in the treatment of long-bone fractures in dogs. J Am Vet Med Assoc 166:520–524, 1975.

Hirsh DC, Smith TM. Osteomyelitis in the dog: Microorganisms isolated and susceptibility to antimicrobial agents. J Small Anim Pract 19:679–687, 1978.

Kahn DS, Pritzker KPH. The pathophysiology of bone infection. Clin Orthop Rel Res 96:12–19, 1973.

Kleine LJ. Radiographic diagnosis of epiphyseal plate trauma. J Am Anim Hosp Assoc 7:290–295, 1971.

Lawson DD. The management of fractures in domestic animals. Br Vet J 119:409–421, 492–511, 1963.

Lewis RP, Sutter VG, Finegold SM. Bone infection involving anaerobic bacteria. Medicine 57:279–305, 1978.

Manley PA, Gunn C, Morgan JP. Technique Guide to Fracture Fixation. Davis, CA: Venture Press, 1981.

Marretta SM, Schrader SC. Physeal injuries in the dog: A review of 135 cases. J Am Vet Med Assoc 182:708–710, 1983.

Mero M, Axelson P, Raiha Vainionpaa S, Tormala P. Operating manual for internal fixation of canine and feline cancellous bone and physeal fractures with biodegradable fixation devices. Biofix Vet:1–15, 1989.

Messmer JM, Fierro MF. Radiologic forensic investigation of fatal gunshot wounds. Radiographics 6:457–473, 1986.

Morgan JP. Radiology in Veterinary Orthopedics. Philadelphia: Lea & Febiger, 1972:21–83.

Morgan JP. Radiographic diagnosis of fractures and fracture repair in the dog. Compend Cont Ed 19:1–12, 1978.

Morgan JP. Radiology of Skeletal Disease. Ames, IA: Iowa State University Press, 1981.

Morgan JP (ed). Techniques of Veterinary Radiography, 5th ed. Ames, IA: Iowa State University Press, 1993.

Morshead D. Fracture fixation with Kirschner wires. Compend Cont Ed 4:491–498, 1982.

Newton CD. Surgical management of distal ulnar physeal growth disturbances in dogs. J Am Vet Med Assoc 164:479–487, 1974.

Newton CD, Hohn R. Fracture nonunion resulting from cerclage appliances. J Am Vet Med Assoc 164:503–508, 1974.

Norden CW. Experimental osteomyelitis: 1. A description of the model. J Infect Dis 122:410–418, 1970.

Nunamaker DM. Repair of femoral head and neck fractures by interfragmentary compression. J Am Vet Med Assoc 162:569–572, 1973.

O'Brien TR. Developmental deformities due to arrested epiphyseal growth. Vet Clin North Am 1:441–454, 1971.

O'Brien TR, Morgan JP, Suter PF. Epiphyseal plate injury in the dog: A radiographic study of growth disturbance in the forelimb. J Small Anim Pract 12:19–35, 1971.

Pavlov H, Torg JS, Hersh A, Freiberger RH. The roentgen examination of runners' injuries. Radiographics 1:17–34, 1981.

Payne JT. Identifying and managing intra-articular fractures in dogs and cats. Compend Cont Ed 10:974–980, 1993.

Pond MJ. Management of avulsion and articular fractures. Vet Clin North Am 2:241–249, 1975.

Ramadan RO, Vaughan LC. Premature closure of the distal ulnar growth plate in dogs: A review of 58 cases. J Small Anim Pract 19:647–667, 1978.

Rang M. The Growth Plate and Its Disorders. Edinburgh and London: E & S Livingstone Ltd., 1975.

Rhinelander FW. Some aspects of the microcirculation of healing bone. Clin Orthop 40:12–16, 1965.

Rhinelander FW. The normal microcirculation of diaphyseal cortex and its response to fracture. J Bone Joint Surg [Am] 50:784–800, 1968.

Rhinelander FW. Tibial blood supply in relation to fracture healing. Clin Orthop 105:34–81, 1974.

Rhinelander FW, Baragry RA. Microangiography in bone healing: I. Undisplaced closed fractures. J Bone Joint Surg [Am] 44:1273–1298, 1962.

Rhinelander FW, Phillips RS, Steel WM, Beer JC. Microangiography in bone healing: II. Displaced closed fractures. J Bone Joint Surg [Am] 50:643–662, 1968.

Richardson ML, Kilcoyne RF, Mayo KA, Lamont JG, Hastrup W. Radiographic evaluation of modern orthopedic fixation devices. Radiographics 7:685–701, 1987.

Rogers LF. The radiography of epiphyseal injuries. Radiology 96:289–299, 1970.

Salter RB, Harris WR. Injuries involving the epiphyseal plate. J Bone Joint Surg [Am] 45:587–622, 1963.

Schwarz PD, Schrader SC. Ulnar fracture and dislocation of the proximal radial epiphysis (Monteggia lesion) in the dog and cat: A review of 28 cases. J Am Vet Med Assoc 185:190–194, 1984.

Sisk TD. External fixation: Historical review, advantages, disadvantages, complications, and indications. Clin Orthop 180:15–22, 1983.

Slone RM, Heare MM, Vander Griend RA, Montgomery WJ. Orthopedic fixation devices. Radiographics 11:823–847, 1991.

Sumner-Smith G, Cawley AJ. Nonunion fractures in the dog. J Small Anim Pract 11:311–325, 1970.

Trueta J. The role of the vessels in osteogenesis. J Bone Joint Surg [Br] 45:402–418, 1963.

Urist MR, Johnson RW. The healing of fractures in man under clinical conditions. J Bone Joint Surg [Am] 25:375–426, 1943.

Vaughan LC, France C. Abnormalities of the volar and plantar sesamoid bones in Rottweilers. J Small Anim Pract 27:551–558, 1986.

Waldvogel FA, Mesdoff G, Swartz MN. Osteomyelitis: A review of clinical features, therapeutic considerations and unusual aspects. N Engl J Med 282:198–206, 260–266, 316–322, 1970.

Walker RD, Richardson DC, Bryant MJ, Draper CS. Anaerobic bacteria associated with osteomyelitis in domestic animals. J Am Vet Med Assoc 182:814–816, 1983.

Whiteside LA, Kouske O, Lesker P, Reynolds FC. The acute effects of periosteal stripping and medullary reaming on regional bone blood flow. Clin Orthop 131:266–272, 1978.

Wilson JW, Rhinelander FW, Stewart CL. Vascularization of cancellous chip bone grafts. Am J Vet Res 46:1691–1699, 1985.

Withrow SJ, Holmberg DL. Use of full cerclage wires in the fixation of 18 consecutive long-bone fractures in small animals. J Am Anim Hosp Assoc 13:735–743, 1977.

Case Studies

Scapula

Signalment: 1 year, female, intact, Irish Setter

History: dog injured 1 week ago

Physical Examination: lame left forelimb with crepitus in left shoulder

Radiographic Examination: left shoulder

Study at time of entry (single lateral view) (day 1):

1. multiple fractures of neck of left scapula with an articular component
2. fracture of the spine of the scapula
3. displacement of fracture fragments
4. radiographically normal proximal humerus

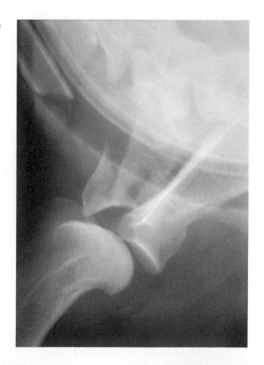

Postoperative study (day 1):

1. reduction using a K wire with a single screw for stabilization
2. excellent apposition and alignment of the articular fragment
3. tension-band device reduces the greater tubercle osteotomy
4. hemicerclage wire reduces the acromion osteotomy

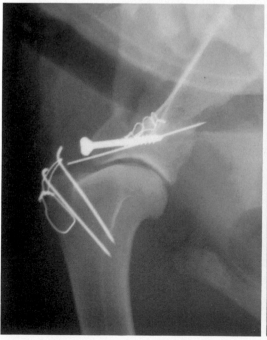
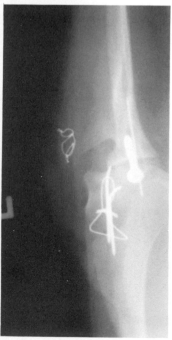

Postoperative study (day 38):

1. orthopedic materiel as before
2. fracture lines poorly visualized
3. poorly organized callus forming around the fracture site (arrows)
4. apposition and alignment of fragments as before

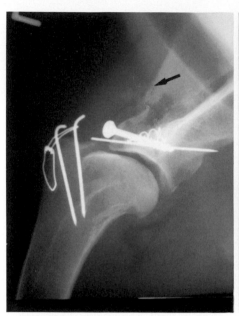 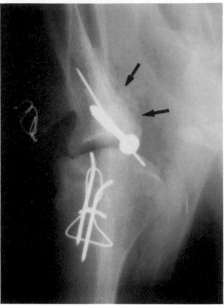

Comments on radiographic findings:

1. note that the original radiographs were made with the patient anesthetized with endotracheal tube and esophageal stethoscope in position, at the time of surgical correction of the lesion
2. callus formation around fractures in a flat bone with poorly developed periosteum is much less prominent and likely to give the appearance of "fluffy" periosteal new bone (arrow)
3. note the continued separation of the spine from the scapular body
4. an important feature to evaluate radiographically is the degree of reduction of the articular fragment

Summary:

The most important feature in this fracture is the articular component and the recognition that anatomic reduction must be obtained. Surgical exposure was obtained by osteotomy of the humeral tubercle to release the supraspinatus muscle and osteotomy of the acromion process to release the acromion deltoid muscle. Joint congruity was reestablished by placement of a K wire. After the contour of the glenoid cavity was restored, the fragment was stabilized through placement of a lag screw. The osteotomy sites were then reduced using two K wires and a figure-8 tension band on the greater tubercle, and two loops of orthopedic wire.

This is an example of a difficult fracture that the surgeon handled in an appropriate manner.

Signalment: 1 year, male, intact, Doberman Pinscher

History: unknown

Physical Examination: dog lame on the right forelimb and painful on palpation of the scapula

Radiographic Examination: right shoulder

Study at time of entry (day 1):

1. comminuted fracture in the distal third of the scapula with separation of the spine distally
2. at least one butterfly fragment cranially
3. fracture of the acromion process
4. 1–2-cm overriding of fracture fragments
5. radiographically normal shoulder joint and first ribs
6. no radiographic evidence of hemothorax or pneumothorax

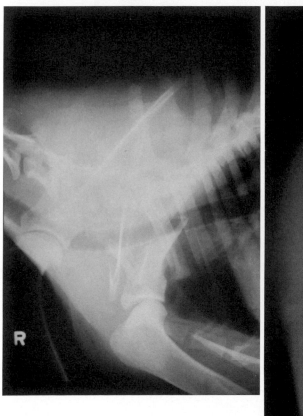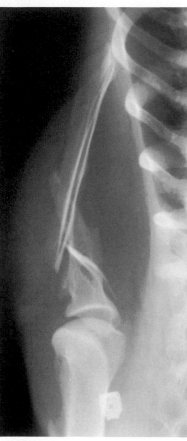

Postoperative study (day 1):

1. reduction using several wire loops
2. stabilization of the fracture using a small bone plate with three screws proximally and three screws distally

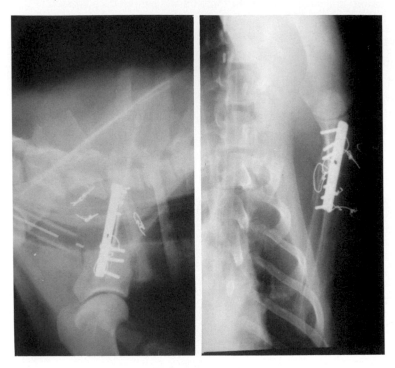

Comments on radiographic findings:

1. a bright light is needed to evaluate completely the fractures of the acromion process
2. examination of the cranial thorax on a patient with scapular fractures may reveal injury to the first ribs or lungs

Summary:

This is a good study of a nonarticular scapular fracture. The spine has broken free and the blade is fractured with overriding. Fractures of this type need to be reduced because the overriding is painful. Injury of the brachial plexus is a possibility, especially if there is complete abduction of the limb at the time of injury. A careful neurologic examination is indicated. In addition, the scapular nerve, which courses over the scapular neck, needs to be examined for injury and carefully protected during surgery.

The surgeon reduced the fracture of the blade by wiring the fragments and then wiring the spine back into position. A wire was used to attach the acromion fragment. The wires were used to reduce the fragments and were not intended as the sole means of fixation. A straight plate was chosen because the caudal angulation of the spine provides a flat surface for its placement. If a tubular plate is available, it can be placed caudal to the spine and positioned satisfactorily with smaller screws. The resulting rigid stabilization is important for controlling pain as well as hastening healing.

The orthopedic wires used here are made with an eyed loop. The end of the wire is drawn through this loop with a special wire tightener, bent at an acute angle, and cut off.

Humerus

Signalment: 6 months old, female, intact, English Bulldog

History: acute lameness in right thoracic limb

Physical Examination: nonweight bearing on right thoracic limb, minimal soft tissue swelling

Radiographic Examination: right elbow

Study at time of entry (day 1):

1. oblique articular fracture distal humerus (Salter-Harris type IV) (arrows)
2. proximal displacement of lateral portion of the condyle (capitulum)
3. growth plates normal and open, as expected for age
4. soft tissue swelling

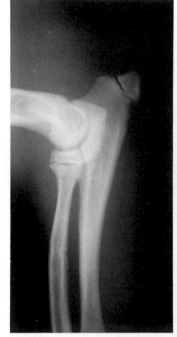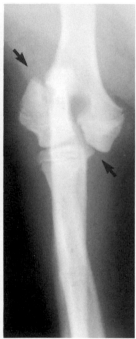

Postreduction study (day 3):

1. transcondylar lag screw inserted medial to lateral
2. single K wire crosses fracture line
3. misplaced tension-band device used to replace olecranon osteotomy
4. radiopaque soft tissue suture material

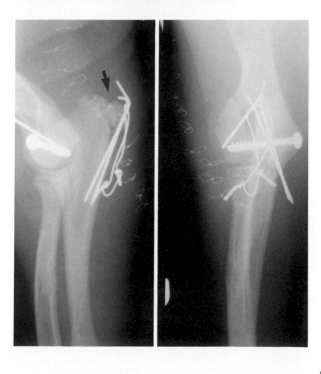

Comments on radiographic findings:

1. note that the lateral radiograph could easily have been misdiagnosed as radiographically normal—the craniocaudal view is essential in diagnosis of elbow injury of this type
2. note that the osteotomy site is not easily recognized, suggesting that it was made through the caudal tip of the apophyseal growth center

Summary:

This most common lateral condylar fracture appears to be a low-energy fracture that was not handled in the best manner. The intercondylar lag screw was introduced from the medial epicondyle, whereas introduction from the lateral epicondyle is more customary. However, good position of the fragment was obtained. Improved placement of the interfragmentary wire could have been achieved by exposing the surface of the fracture fragment and then driving the wire through the center of the condylar fragment in a proximal-to-distal direction. The wire is then reinserted proximally until it reaches the endosteal surface of the medial humeral cortex. Then, the lag screw could be tightened so the articular surfaces are congruent.

The osteotomy (arrow) should have been placed distal to the growth plate so that it included a greater amount of bony tissue. This provides for more stable replacement of the osteotomized fragment. By predrilling the olecranon before transection, reduction is accomplished with the fragments replaced in an anatomic position.

Bulldogs and other members of chondrodystrophoid breeds bring more acutely into focus problems associated with placement of K wires to reduce the osteotomized fragment. The proximal ulna and olecranon in these dogs curves markedly in a lateral direction (medial concavity), creating considerable bowing. The osteotomy wires must be introduced medially to the tip of the olecranon (see Fig. 8-2) and angled laterally so that they remain within the olecranon as they pass distally.

Despite these problems of reduction and fixation, this fracture healed because of the young age of the patient, the brief time necessary for fracture healing, and the light body weight of the patient. However, healing probably will be associated with the unnecessary development of degenerative joint disease, and it is likely that the elbow will have limited range of motion after healing.

The osteotomy replacement pins eventually must be removed in this patient because of their penetration into soft tissues medially.

Signalment: 3 months old, female, intact, Labrador Retriever

History: owner reported that the puppy was running in a field and stepped in a hole, and became acutely nonweight bearing on the right forelimb

Physical Examination: nonweight bearing on right forelimb, crepitus and swelling noted at the distal humerus

Radiographic Examination: right elbow

Study at time of entry (day 1):

1. type 4 Salter-Harris fracture of distal humerus with separation of the lateral condyle (arrows)
2. minimal displacement of the fracture fragment
3. remaining growth plates are radiographically normal and as expected for age

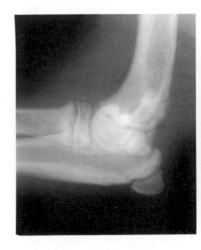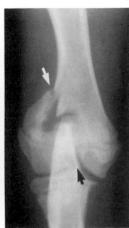

Postreduction study (day 1):

1. reduction of condylar component using a transcondylar, fully threaded cancellous screw
2. wire placed adjacent to the screw used to prevent rotation of the fragment
3. wire passed through the lateral metaphysis into the shaft

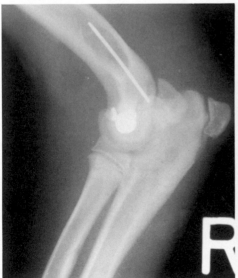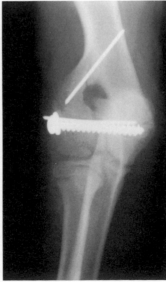

Postreduction study (11 months):

1. healing of the fracture
2. good position of the fragment
3. metallic devices remain in position
4. soft tissue calcification at the site of screw placement (arrows)

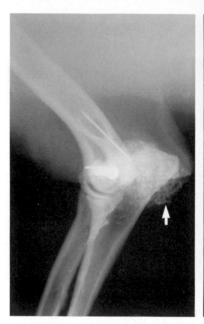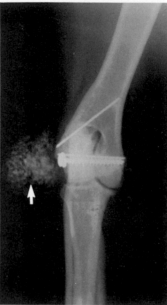

Comments on radiographic findings:

1. fracture cannot be identified on the lateral view
2. note that the fracture line passes into the joint space (arrow)
3. fracture healing occurred early with minimal callus because of good stability of the fragment in a young patient
4. dystrophic calcification (arrows) posttrauma

Summary:

This is a low-energy fracture treated with a very acceptable technique. The fracture is in a young patient and the reduction was rigid and anatomic. Healing will occur early and with only minimal callus formation. Actually, no callus was seen radiographically during the healing process. The effort to approximate the articular surfaces was successful. The transcondylar screw is placed ideally in that it is centrally located, avoiding distraction and misplacement of the condylar fragment. The small IM wire was introduced from the lateral epicondyle and seated into the medial cortex of the distal metaphysis. Insertion in this manner provides for improved rotational stability of the condylar fragment.

The posttraumatic soft tissue calcification is uncommon. It is most likely secondary to hemorrhage at the time of trauma and unrelated to fracture healing. With maturity, this calcified hematoma may resorb and probably does not require surgical intervention.

Signalment: 9 months old, female, spayed, mixed breed Labrador Retriever

History: dog missing from home for 24 hours; owner found dog after it apparently had been struck by a car on the previous day; dog unable to stand—owner noted swelling over right stifle and left thoracic limb, and difficulty with breathing

Physical Examination: unable to stand; crepitus noted on physical examination of the distal femur on the right and distal humerus on the left; heart and lungs were normal on auscultation

Radiographic Examination: left humerus

Study at time of entry (single lateral view) (day 1):

1. oblique fracture at the junction of the middle and distal thirds of the humerus
2. marked displacement of fracture fragments with overriding
3. large single "butterfly" fragment involving > 50% of the diameter of the bone (arrow)
4. both shoulder and elbow joints are radiographically normal
5. growth plates are normal and appear open, as expected in a dog of this age

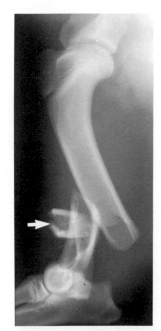

Postreduction study (day 1):

1. reduction and fixation of humeral fracture using a contoured bone plate and threaded cancellous screws with three screws seated within the proximal fragment, two screws at the fracture site in the region of the butterfly fragment, and one screw within the distal fragment
2. type 1 external fixator with one fixation pin within the proximal fragment and one fixation pin within the distal fragment (1/1)
3. lucent zone within the proximal metaphysis at harvest site for cancellous bone graft (arrow)

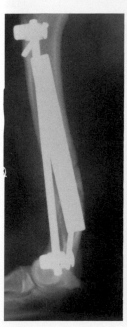
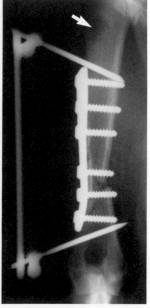

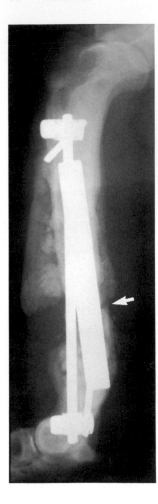

Postreduction study (single lateral view) (day 24):

1. fixation devices remain as noted earlier
2. massive callus forming proximally and distally with no evidence of bridging callus
3. fracture line still evident (arrow)

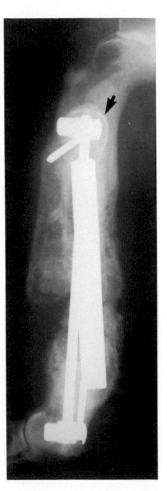

Postreduction study (single lateral view) (day 42):

1. fixation devices remain as noted before
2. massive callus forming proximally and distally with beginning evidence of callus bridging the fracture caudally
3. circular opacity resulting from bandage around proximal fixation pin (arrow)

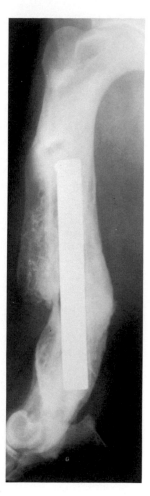

Postreduction study (single lateral view) (day 106):

1. removal of the external K-E device with bone plate and screws as before
2. callus is bridging caudally
3. callus is maturing with margins becoming smooth and a uniform density replacing the "lacy" pattern associated with the original laydown of the callus
4. early filling of pin track
5. filling cancellous graft harvest site

Postreduction study (single lateral view) (day 150):

1. bone plate and screws remain as noted earlier
2. callus is bridging cranially and caudally
3. callus is maturing further with margins becoming smooth and uniformly dense
4. continued filling of pin track

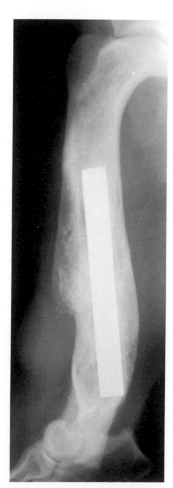

Comments on radiographic findings:

1. in the healing of a fracture of this type in a skeletally immature patient with extensive periosteal stripping, a heavy callus forms subperiosteally—the lack of callus formation from the butterfly fragment at the fracture site indicates loss of periosteum and loss of blood supply to this fragment
2. separated fragment is resorbed into bridging callus
3. exuberant callus formation of this type is suggestive of osteomyelitis; however, a more plausible explanation of callus formation in this patient is associated with original periosteal tearing
4. a final radiographic study (not shown) was made 15 months after the original injury and showed continued modeling of the fracture callus so that the cortices and medullary cavity could once again be identified

Summary:

Our attention was focused on the humeral fracture and its healing. The femoral fracture was physeal, was reduced anatomically, and healed quickly and without difficulty. The humerus is a large, heavy bone, provided with considerable soft tissue protection proximally. Distal fracture with the type of fragmentation as noted here is typical for the larger breeds. The surgeon chose not to perform interfragmentary compression separately, attempting instead to accomplish this with the screws used to attach the bone plate. Selection of a bone plate that was too short resulted in placement of screws dangerously close to the fracture line and provided a less-than-satisfactory attachment of the plate distally. Plating of itself should be an adequate method of fracture reduction and fixation and should not need additional hardware. Use of the external K-E device indicates that the surgeon feared potential failure of the plate and compensated for this possible instability at the fracture site. With the K-E device, the distal fixation pin ordinarily would have been intracondylar, with an additional fixation pin placed in the distal fragment. Two pins inserted proximally would have provided additional stability. The surgeon used a plating technique that was less than desirable and provided no interfragmentary compression. The additional K-E device also appears to be somewhat inadequate.

This common fracture needs to be treated with uncommon care. The fragment(s) should be well secured and a plate of appropriate length selected. This case shows a prolonged and protracted healing period that could have been avoided by use of more secure plating. However, it is noted that the plate remained solidly in position and that satisfactory healing occurred.

Signalment: 7 months old, male, intact, Golden Retriever

History: dog was being transported without restraint in the back of a pickup truck—dog jumped from truck to chase a cat

Physical Examination: nonweight bearing on right thoracic limb; marked swelling in region of elbow, associated right mandibular fracture

Radiographic Examination: right elbow

Study at time of entry (day 1):

1. distal humeral fracture involving the distal physis medially (Salter-Harris type II) with minimal comminution
2. marked overriding of fracture fragments
3. shoulder and elbow joints appear radiographically normal
4. growth plates appear open as expected for a puppy of this age

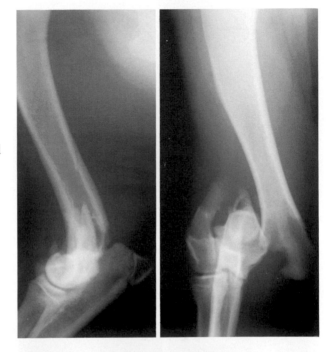

First postreduction study (day 1):

1. full-length IM pins placed in a crossing manner (modified Rush pin technique)
2. hemicerclage wire encircles the lateral pin
3. tension-band device repositions the small osteotomy fragment
4. failure of medial wire associated with the tension-band device to remain within the ulna (arrow)
5. hole drilled for wire in tension-band device is cortical and shallow
6. postsurgical soft tissue gas

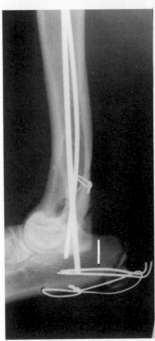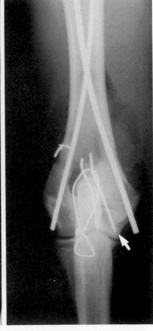

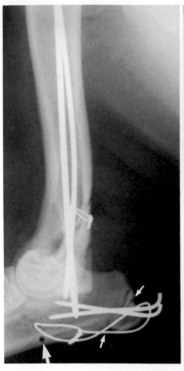
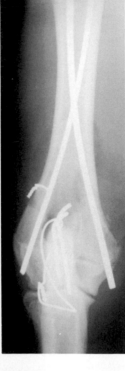

Second post-reduction study (day 1):

1. major fixation devices remain essentially as before
2. original tension-band device replaced using heavier wires, wires now remain within the medullary cavity, and a new drill hole is used for the tension-band wire (arrow)
3. postsurgical soft tissue gas

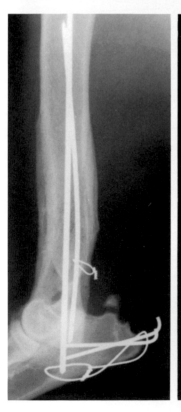
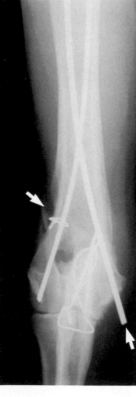

Postreduction study (day 100):

1. IM pins remain in position with minimal backing-out of medial pin (arrow)
2. 3-cm increase in length of humerus due to bone growth, resulting in the lateral pin being buried within the lateral condyle
3. medial pin has slipped distally and protrudes within the soft tissues
4. hemicerclage wire remains in position
5. smooth, mature-appearing callus formation extends to the mid-diaphysis
6. fracture lines not identified
7. tension-band device remains as before
8. avulsion fragment from proximal olecranon

Comments on radiographic findings:

1. osteotomy site can be located more distally within the olecranon (line) instead of at the growth plate (small arrows)
2. difficult to determine on original study whether there is fragmentation at the fracture site; although suggested, later studies prove that this is not the case
3. important to note a change in location of metallic devices at the time of healing—one pin becomes buried due to bone growth, possibly because of increased contact with the medial endosteal surface, whereas the other backs out slightly
4. long callus formation is caused by periosteal stripping that extends to the midshaft—this is an expected pattern of callus formation secondary to a hyperactive cellular population of osteoblasts beneath the periosteum in the immature skeleton
5. break in callus formation at site of cerclage wire suggests periosteal tearing at that point associated with surgical placement of the wire, resulting in delayed callus formation (arrow)
6. the avulsion fragment from the proximal olecranon process actually appears to originate slightly distal to its location; this probably is associated with the trauma of the repeated surgical procedures performed in this region
7. it is important to evaluate bone growth radiographically because the original fracture was physeal in character; radiographs of the opposite limb can be helpful in determining normal growth

Summary:

This fracture is unusual because it is without an articular component. The osteotomy technique used to gain surgical exposure separated a very small bony fragment. It is preferable to make the osteotomy site more distal. Two flexible IM pins placed from the medial and lateral epicondylar regions were used in a modified Rush pin technique. The hemicerclage wire was used to stabilize the lateral supracondylar crest against one pin. Unfortunately, the tension-band K wires originally used were small and improperly placed. The surgeon chose to redo the tension band device using larger K wires properly inserted within the ulna and a new drill hole that was more deeply seated for the figure-8 tension-band wire. The last study showed good healing with an exuberant callus secondary to periosteal stripping. The lateral pin became buried as a result of the growth of the humerus. Why the medial pin moved with the growing bone and remained protruding is not clear. The origin of the small avulsed bone fragment is uncertain, but probably is related to surgical trauma.

This is a good technique to use in treatment of fractures of this type, and clearly shows the value of the modified Rush pin technique. Use of an external fixator or bone plate would have been difficult because of the small size of the distal fragments. A minor technical problem is noted relative to the olecranon osteotomy. An understanding of the anatomic features of the proximal ulna is important. If the wires are driven straight, ignoring the curvature of the proximal ulna, they will penetrate the medial cortex, offering decreased holding power and causing irritation of the soft tissues. The pins must be aimed laterally to remain within the ulna.

CASE 5

Signalment: 2 years old, male, castrated, domestic short-haired cat

History: unknown trauma 48 hours previously

Physical Examination: nonweight bearing on left thoracic limb; marked soft tissue swelling in region of elbow; crepitus near elbow

Radiographic Examination: left elbow

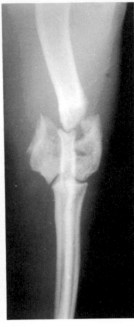

Study at time of entry (day 1):

1. "T" type articular fracture of distal humerus with subluxation
2. separation of the condylar parts
3. stage of physeal closure suggests an age of 12 to 14 months
4. shoulder joint appears radiographically normal

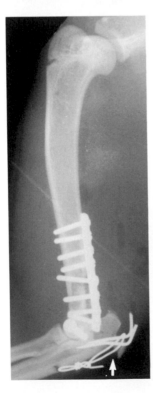
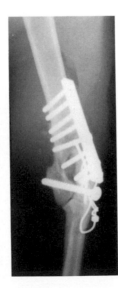

Postreduction study (day 2):

1. placement of transcondylar, fully threaded cortical screw
2. placement of eight-hole contoured bone plate
3. use of a figure-8 tension-band device to reduce olecranon osteotomy fragment
4. postsurgical soft tissue gas

Comments on radiographic findings:

1. difficult to interpret the character of the distal humeral fragment as visualized through the olecranon process on the craniocaudal view
2. it is wise to note the stage of bone growth radiographically because owners frequently forget the age of their pets

Summary:
In the cat, the transcondylar screw can be a regular cortical screw used in a lag fashion to achieve the necessary fixation. A caudolaterally positioned plate is not as easy to use because of interference from the lateral supracondylar crest. Note the apparent elevation of the plate at this site. Although there are four screws proximally, there appear to be two screws distally that are supporting and two screws that are lagging.

In the cat it is possible to cut the attachment of the triceps muscle adjacent to the olecranon and reattach it with the placement of nonabsorbable suture. However, if the olecranon is to be osteotomized, creation of a larger fragment would furnish better seating for the pins and easier replacement (arrow).

An alternative to this satisfactory form of plating would be the use of small-diameter wires in a modified Rush technique.

CASE 6

Signalment: 1 year old, male, castrated, domestic short-haired cat

History: came home yesterday dragging right front leg

Physical Examination: nonweight bearing on right forelimb, crepitus in elbow

Radiographic Examination: right elbow (humerus)

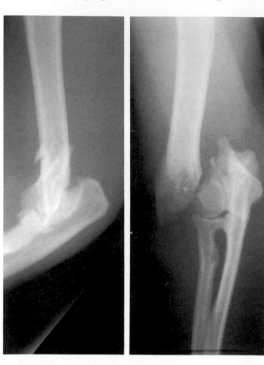

Study at time of entry (day 1):

1. acute oblique fracture of distal humerus, nonarticular with minimal fragmentation
2. severely overriding fragments
3. elbow joint appears radiographically normal, although the distal portion of the limb is internally rotated and difficult to evaluate on both views
4. growth plates in distal humerus and proximal radius and ulna are closed

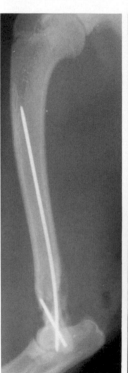
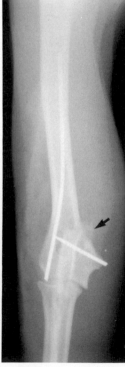

Postreduction study (day 3):

1. single IM pin introduced from the lateral condyle into the shaft of the humerus from the lateral portion of the condyle
2. single wire introduced transversely across the fracture line from the medial portion of the condyle
3. misplaced fragment medially at fracture site that appears as callus formation (arrow)
4. postsurgical soft tissue gas
5. growth plates in proximal humerus, as expected for age

Postreduction study (day 90):

1. fixation devices remain essentially as before
2. fracture line not identified
3. callus is forming an extraperiosteal pattern medially, indicating a periosteal tear from the proximal fragment
4. periarticular lipping (arrow)

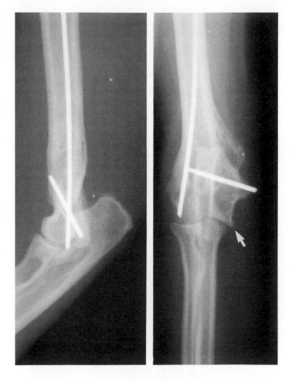

Comments on radiographic findings:

1. exuberant callus just proximal to the medial condyle represents the extent of periosteal tearing in a young patient
2. periarticular lipping noted medially is probably posttraumatic in origin

Summary:

The IM pin was introduced from the lateral portion of the condyle using a modified Rush pin technique, achieving excellent reduction. The medial fragment was stabilized by introduction of a short wire that just crossed the fracture line. The modified Rush pin technique using a single thin pin introduced in conjunction with a second, shorter pin to stabilize the medial condylar portion is acceptable for this type of fracture. The pins are buried and there is nothing to remove later. It would have been possible to place the modified Rush pin from the medial condyle.

The potential for failure in a fracture of this type is high unless internal fixation is used to obtain reduction and stabilization. Note that the surgeon did not cut the olecranon to gain surgical exposure. Usually in the cat, the tendinous insertion of the triceps muscle is cut to achieve exposure at the time of surgery. The unique morphology of the distal humerus in the cat provided the surgeon with a wider latitude in selection of the pin placement technique.

Because the activities of the cat can be more successfully controlled, the additional supporting strength required in the dog is not necessary.

CASE 7 **Signalment:** 6 years old, female, spayed, domestic short-haired cat

History: cat returned home in the morning limping

Physical Examination: crepitus right elbow, laceration in right axilla, subcutaneous emphysema, mild pneumothorax determined radiographically

Radiographic Examination: right elbow

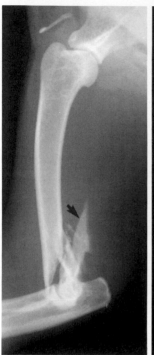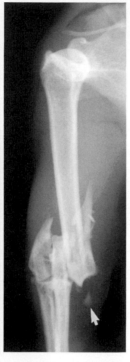

Study at time of entry (day 1):

1. badly comminuted "Y" type fracture of the distal humerus with large butterfly fragment (black arrow)
2. disruption of articular components with overriding
3. small fragments medially and distally (white arrow)
4. shoulder joint appears radiographically normal
5. subcutaneous emphysema

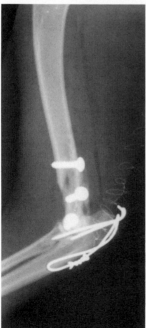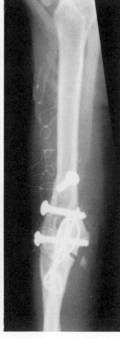

Postreduction study (day 2):

1. placement of two interfragmentary screws
2. placement of transcondylar screw
3. figure-8 tension-band device reduces the osteotomy fragment
4. wires enter elbow joint
5. subcutaneous emphysema

Postreduction study (day 45):

1. screws remain as before
2. fracture lines remain partially visualized
3. only minimal callus formed
4. withdrawal of one intra-articular wire
5. retention of one intra-articular wire

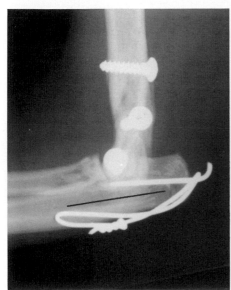 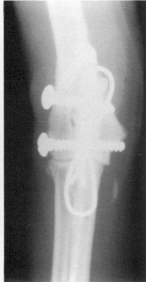

Comments on radiographic findings:

1. exact nature of distal fragments is difficult to ascertain because of the overlying shadows of the olecranon; origin of the small fragments in the soft tissue medially is not certain
2. status of the ulnar trochlear notch is not clear—are some of the fragments from the coronoid processes?
3. it is important to note the status of the adjacent joint in a fracture patient
4. if fracture stabilization is good the amount of callus is minimal, as in this patient
5. origin of the small bony fragment medial to the proximal radius–ulna is difficult to determine; fragment is seen to be resorbing and has no long-term clinical significance

Summary:
A typical distal condylar fracture was handled quite adequately with a transcondylar screw and two lag screws, which brought about a nice anatomic approximation of the fragments. It is unfortunate that the wires used to fix the osteotomized olecranon were inserted into the elbow joint. Removal of the worst offender lessened the damage to the joint. A study of the radiograph shows the proper track for these pins (black line). Surgeons need to be aware of this clinical application of radiographic anatomy.

Once again, the surgeon is relying on being able to confine the cat, which prevents stressing a fracture treated in this fashion.

Signalment: 7 years old, male, intact, Cocker Spaniel

History: owner had just acquired the dog 3 days ago; previous evening he kicked the dog in an effort to prevent progression of a dog fight, and patient landed on a front porch step some distance away; dog was immediately nonweight bearing on right thoracic limb

Physical Examination: nonweight bearing on right thoracic limb; crepitus could not be identified

Radiographic Examination: right humerus

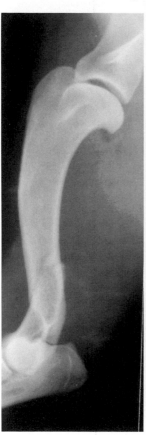 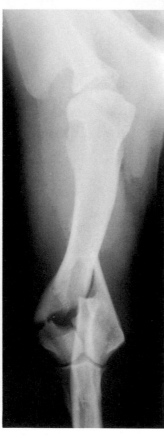

Study at time of entry (day 1):

1. "Y" type fracture of distal humerus with articular component
2. moderate separation of fracture fragments
3. minimal soft tissue injury and fragment displacement suggests a low-energy trauma
4. shoulder joint radiographically normal

Postreduction study (day 1):

1. reduction of condylar component using a transcondylar, fully threaded cancellous screw
2. single smooth IM pin with tip (arrow) passing through the cortex of the medial epicondyle (line)
3. tension-band reduction of osteotomized olecranon fragment

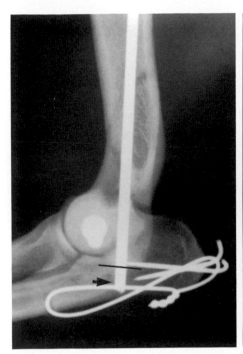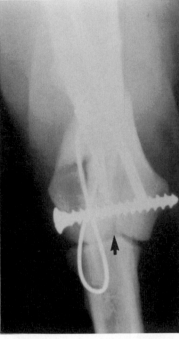

Comments on radiographic findings:

1. note that apposition and alignment of fracture fragments is almost anatomic
2. minimal "stair-step" malalignment of distal humeral articular surface, lacking perfect congruency of articular surfaces (arrow)
3. anatomic replacement of osteotomized fragment, with osteotomy site not identified on the radiographs

Summary:
This is a commonly occurring low-energy fracture with a very acceptable technique of treatment. Every effort was made to approximate the articular surfaces. The transcondylar screw is ideally placed in that it is properly located so as to be in the center of the condylar fracture faces. Even though placement appeared satisfactory, minimal distraction and misplacement of the condylar fragment occurred. The IM pin was introduced distally from the medial epicondyle, driven proximally, and seated deeply into the proximal humerus. Insertion in this manner provides for improved stability of the medial condylar fragment. Still, a more common method of introduction would be in a normograde direction with the tip driven into the medial condyle. Because the IM pin protrudes distally, it must be guarded during the healing phase and must be removed after fracture healing.

The transolecranon osteotomy is essential to obtain visual exposure for reduction and fixation. Placement of the figure-8 tension-band wires and K wires has been done properly, although the wires are noted to be short.

Signalment: 5 years old, male, castrated, Border Collie

History: struck by a car, unable to walk using right forelimb

Physical Examination: nonweight bearing on right forelimb, marked crepitus in region of elbow

Radiographic Examination: right elbow

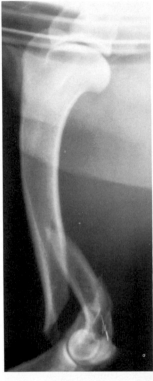
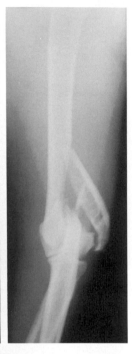

Study at time of entry (day 1):

1. acute, long, oblique nonarticular fracture of distal humerus
2. large butterfly fragment including the lateral supracondylar crest
3. overriding fracture fragments
4. shoulder and elbow joints appear radiographically normal

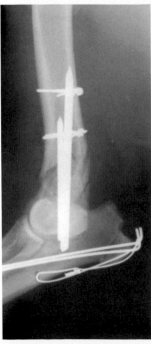
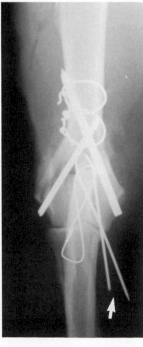

First postreduction study (day 1):

1. single, short IM pin introduced in a retrograde manner from the medial epicondyle penetrates the lateral cortical shaft
2. second, smaller IM pin introduced from the lateral epicondyle placed in a crossing fashion with the tip seated into the medial cortex
3. two cerclage wires position the proximal portion of the butterfly fragment
4. cancellous graft used
5. tension-band device repositions the osteotomy fragment
6. failure of wires associated with the tension-band device to remain within the ulna (arrow)
7. postsurgical soft tissue gas

Second postreduction study (day 1):

1. major fixation devices remain essentially as before
2. corrected positioning of wires associated with tension-band device

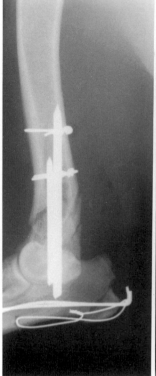
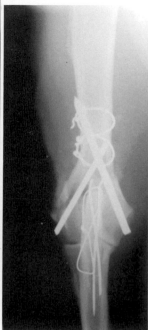

Postreduction study (single craniocaudal view) (day 10):

1. lateral IM pin has backed out (arrow)
2. collapse of the fracture fragments around the larger IM pin
3. cerclage wires remain in position
4. early callus formation along with cancellous graft is noted around the fracture site, and extends to the midshaft of the bone
5. elbow joint space not evaluated on this view

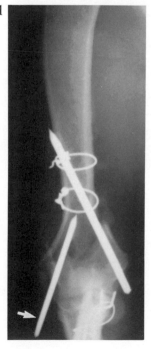

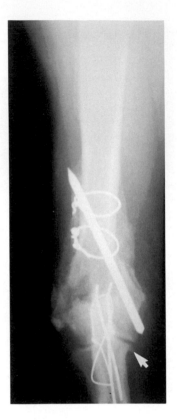

Postreduction study (single craniocaudal view) (day 60):

1. lateral IM pin has been removed
2. persistent collapse of the fragments around the larger IM pin with resulting varus deformity
3. cerclage wires remain in position
4. more mature callus formation is noted around the fracture site, partially obliterating the fracture lines
5. periarticular spurring in region of medial coronoid process (arrow) suggesting elbow joint injury and early radiographic changes of joint disease

Comments on radiographic findings:

1. note use of the "eyed" cerclage wires
2. malposition of the osteotomy fragment suggests that holes were not predrilled
3. pattern of callus formation was exuberant and extended to midshaft due to instability at fracture site—the pattern of callus formation plus the use of cancellous graft makes radiographic evaluation of the fracture difficult on the 10-day study (is it unstable? is it infected?)
4. periarticular lipping suggests possibility of early bony changes associated with arthrosis formation—cause of the arthrosis not fully understood because the original injury was not articular and the joint should not have been invaded surgically
5. postreduction craniocaudal radiographs were difficult to make with the patient in sternal recumbency and the thoracic limb fully extended; this view is made more easily with the patient in dorsal recumbency with the forelimb pulled caudally next to the thorax and the humerus positioned parallel to the film—increased object–film distance creates only minimal loss of detail and is not a contraindication
6. status of the elbow joint is not as well shown in this patient as it should be

Summary:

This unusual fracture has a large butterfly fragment that includes the lateral supracondylar crest. The osteotomy cut used to gain surgical exposure was poorly located. Two cerclage wires were used to stabilize the large butterfly fragment; however, they provide stability only to the proximal portion of the fragment. The IM pin positioned from the medial epicondyle protrudes through the lateral cortex, with good reduction and stabilization of the medial fragment. The lateral fragment was stabilized by placement of a second, smaller IM pin from the lateral epicondyle using a cross-pinning technique. The smaller pin engages but does not penetrate the medial cortex. Replacement of the osteotomy fragment was done poorly, with the pins protruding medially. A second radiographic study was performed, showing the correct position of the osteotomy wires.

Unfortunately, the small IM pin backed out laterally, having lost its hold on the endosteal surface. This permitted motion of the distal fragment, which drifted medially with impaction and medial angulation. This in turn forced the tip of the cut end of the lateral IM pin further into the soft tissues. It was hoped that healing had progressed sufficiently that with removal of the lateral pin healing would proceed satisfactorily. Unfortunately, this did not happen, and further motion of the distal fragment occurred. Because of instability, an exuberant callus was required to heal the fracture.

The technique selected for fracture repair has proven satisfactory in selected cases. Cross pinning should have met with success; however, the failure of the smaller pin to penetrate the cortex permitted it to disengage. In addition, the patient was released and permitted extremely active and unsupervised exercise. It was optimistic to assume that the large butterfly fragment would remain stabilized only through the use of the cerclage wires, which were positioned so far proximally.

A modified Rush pin technique might have provided more stabilization, except that in this patient, the large lateral butterfly fragment extended distally to the condyle. Therefore, it is necessary to consider stabilizing this fragment with lag screws. Placement of a lateral pin might have been difficult because of the presence of the lag screws, but placement of the medial pin should have been without difficulty.

CASE 10   **Signalment:** 5 years old, male, intact, Great Dane

History: dog with chronic injury referred with radiographs showing a gunshot injury to the left humeral region

Physical Examination: gunshot wound to left proximal forelimb

Radiographic Examination: left humerus

Study at time of entry (day 50):

1. highly comminuted fracture of lower midshaft humerus
2. marked separation of fracture fragments with bone shortening
3. early callus formation on all fragments
4. metallic fragments within soft tissue (white arrowheads)
5. elbow joint appears radiographically normal

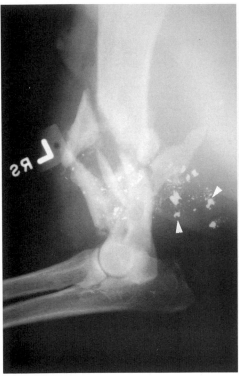 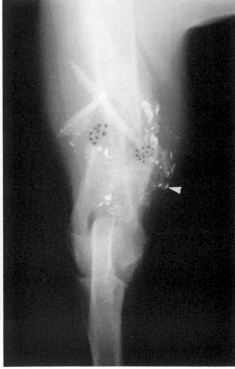

Postoperative study (single lateral view) (day 90):

1. reduction with seven fixation pins stabilized using a plastic mold cranially (numbers 2 and 7 are positive-thread fixation pins)
2. external support in the form of corrugated plastic tube filled with methylmethacrylate
3. lucency around the tip of the fourth screw
4. minimal callus bridges some fragments; however, the major fracture lines still are identified
5. several avascular bone fragments are identified (arrows)
6. metallic fragments as before
7. minimal repositioning of the larger fragments

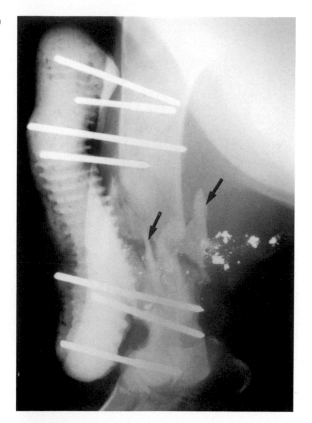

Postoperative study (single lateral view made after dog was returned with proximal pins removed) (day 128):

1. removal of entire device
2. callus formation without bridging of the fracture site (nonunion fracture)
3. avascular fragments clearly identified (sequestra) (arrows)
4. pin tracks easily identified
5. metallic fragments as before

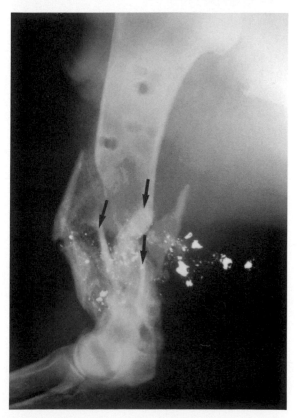

Comments on radiographic findings:

1. note the early callus formation on the bony fragments at the time of the first radiographic study, indicating the chronicity of the injury
2. some fragments remain unchanged after 2 months, suggesting avascularity and possible sequestration with osteomyelitis
3. normal length of the bone can be obtained from radiographs of the opposite limb
4. does the dog have osteomyelitis?—every reason to believe so
5. note the technical errors due to static electricity

Summary:

A 7-week-old fracture due to a gunshot wound was finally stabilized by application of a type I external fixation device. Two of the seven fixation pins have special enlarged threaded ends. These are specially designed for use in cancellous bone. Note that the threads of numbers 2 and 7 are not identified on the lateral view because of obliquity.

The external fixation device was removed 10 weeks after placement with the fracture still ununited. It is assumed that the distal fragment was moving cyclically, acting as a pendulum through the plastic mold to the proximal fragment.

The fracture was in a large dog, and exceptional care had to be taken to ensure stability at the fracture site as well as reduce activity of the patient. The difficulties were compounded by the long delay before fixation.

The patient was lost to follow-up, but the owner was advised to consider amputation of the limb.

Signalment: 3 years old, male, intact, Miniature Poodle

History: owner found dog by side of roadway

Physical Examination: nonweight bearing on right thoracic limb

Radiographic Examination: right humerus

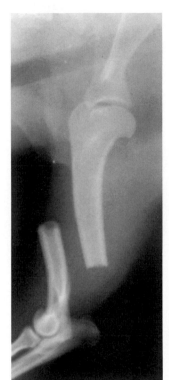

Study at time of entry (single lateral view) (day 1):

1. simple transverse midshaft humeral fracture
2. marked separation of fragments with overriding
3. both shoulder and elbow joints are radiographically normal

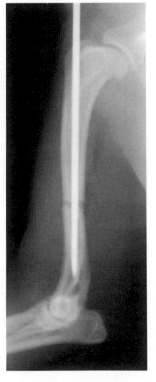

Intraoperative study (single lateral view) (day 1):

1. reduction using a single IM pin with a threaded tip
2. apposition and alignment are anatomical on this single lateral view
3. intraoperative radiograph shows the pin is not yet cut

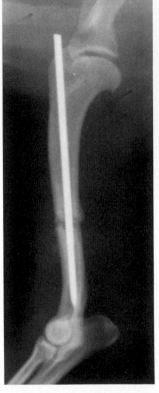

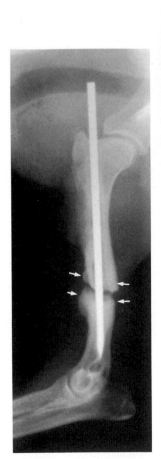

Postreduction study (single lateral view) (day 20):

1. reduction and fixation remain as before
2. apposition and alignment as before
3. no callus formation
4. soft tissue atrophy

Postreduction study (single lateral view) (day 40):

1. reduction and fixation remain as before, with widening at the fracture site
2. apposition and alignment as before
3. minimal callus formation at fracture site
4. sclerotic pattern noted at ends of fracture fragments, with both endosteal and periosteal new bone becoming dense with smooth borders
5. soft tissue atrophy

Comments on radiographic findings:

1. radiographic appearance on the last study is typical of a reactive nonunion fracture with development of an "elephant foot" appearance (arrows)

Summary:

This type of transverse fracture occurs commonly in small breeds, suggesting that muscle attachment in these dogs is insufficiently strong to cause butterfly fragmentation or comminution. Also, these breeds tend to protect the bone after the fracture. The surgeon may be led into thinking that the "simplest" method of reduction and fixation will be satisfactory in handling a "simple" fracture of this type. This is not so! The single IM pin provides no rotational stability. However, it was introduced proximally and passed into the distal fragment in a way that does provide axial stability.

It would have been better to treat the fracture using: (1) a small bone plate, (2) an IM pin and a type I K-E device with two fixation pins placed proximally and a transcondylar fixation pin distally (2/1), or (3) a type I K-E device with three fixation pins placed proximally and three fixation pins placed distally, with an additional pin placed transcondylarly (3/3). Use of a cancellous graft would have assisted greatly in the healing of a fracture of this type, and is almost mandatory.

CASE 12   **Signalment:** 2 years old, female, intact, mixed breed Australian Shepherd

History: owner found dog lame

Physical Examination: nonweight bearing on left forelimb and marked crepitus on examination

Radiographic Examination: left humerus

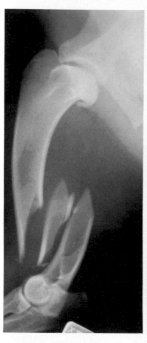 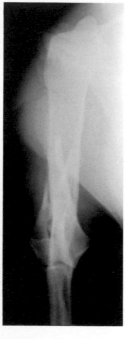

Study at time of entry (day 1):

1. oblique comminuted fracture at the junction of the middle and distal thirds of the humerus
2. marked separation of fragments
3. single large butterfly fragment (> 50% of shaft diameter)
4. chronic arthrosis of shoulder joint with caudal articular fragment suggestive of osteochondritis dissecans
5. elbow joint radiographically normal

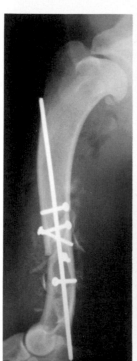 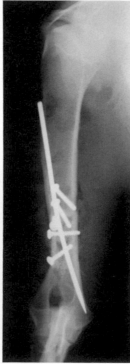

Postreduction study (day 1):

1. interfragmentary compression using multiple small cortical screws
2. small IM pin with a chisel tip placed proximally and driven distally into the medial epicondyle
3. cancellous bone graft placed around the fracture site containing prominent cortical fragments
4. postoperative soft tissue gas noted

Postreduction study (day 15):

1. reduction remains as before, with the single smooth IM pin and multiple small cortical screws
2. fracture line still identified
3. little change in bone graft placed around the fracture site, with persistent prominent cortical fragments
4. minimal callus (periosteal response) forming proximally and distally (arrow)

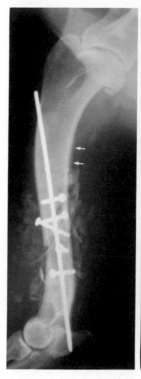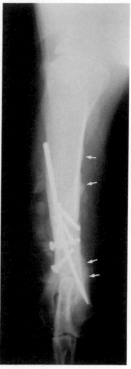

Comments on radiographic findings:

1. presence of the bone graft makes identification of early healing callus and determination of the closure of the fracture line difficult
2. formation of the callus on the proximal fragment suggests stripping of the periosteum, which might be unusual at this age or may suggest surgical trauma

Summary:

Having used screws to achieve interfragmentary compression, as well as using a bone graft, it is unfortunate that the surgeon elected to introduce a small IM pin that would fit around the screws; use of a bone plate would have been a much better choice. Solid fixation of humeral fractures is necessary because 60% of the dog's weight is borne on the thoracic limbs. It would have been acceptable in a fracture of this type to use only an external K-E device with three-pins proximally and three pins distally, with the distal pin threaded and placed transcondylarly into a tapped hole (3/3).

Axial and rotational forces interfered with the healing in this patient, as evidenced by the delay in callus formation noted on the 15-day study. The surgeon at this point must choose between the diagnosis of delayed healing that may lead to nonunion, or stay with this method and wait a longer period of time. Because this is only 15 days postreduction, a diagnosis of potential nonunion is premature. A fracture of this magnitude often does not heal for 8 to 12 weeks.

Signalment: 1 year old, female, spayed, mixed breed dog

History: dog missing from home for 4 days; large soft tissue wound present over the thorax on return, right thoracic limb swollen

Physical Examination: nonweight bearing on right thoracic limb, crepitus in right humerus

Radiographic Examination: right humerus

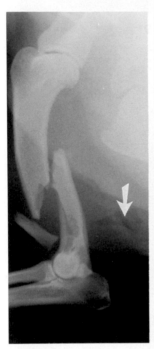
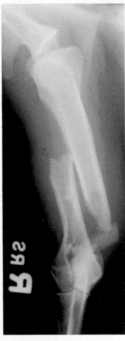

Study at time of entry (day 1):
1. oblique fracture of midshaft humerus with one large butterfly fragment originating from the cranial cortex of the distal fragment (< 50% of shaft diameter)
2. overriding fragments
3. soft tissue gas posttrauma (arrow)
4. both shoulder and elbow joints appear radiographically normal

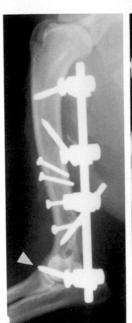
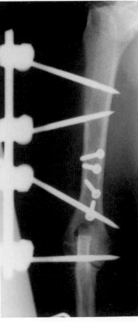

Postreduction study (day 1):
1. two interfragmentary screws used to reduce the butterfly fragment
2. two screws used in a lag technique to achieve reduction of major fracture fragments
3. type 1 external K-E device with fixation pins with threaded tips—two pins proximally and two fixation pins distally (2/2)
4. good apposition and alignment of fracture fragments
5. postsurgical soft tissue gas

Postreduction study (day 28):

1. fixation devices remain essentially as before
2. two distally positioned fixation pins retracted after examination of postoperative radiograph
3. callus is forming within soft tissues away from fracture site
4. new bone response distally around pin holes

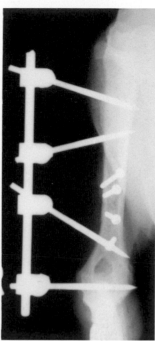

Postreduction study after removal K-E apparatus (single view of elbow) (day 28):

1. condylar destruction seen more clearly (arrows)
2. minimal destruction of articular surfaces
3. removal of the external K-E device

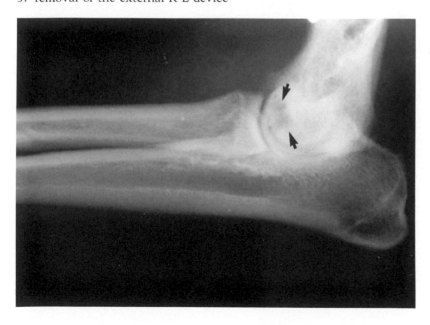

Comments on radiographic findings:

1. note the apparent subluxation of the shoulder joint on the craniocaudal view on the first study—this degree of joint laxity is within normal limits for the shoulder joint
2. the elbow joint cannot be evaluated on the postreduction study (day 28) because the stabilization device overlies the elbow on the lateral view; the craniocaudal view of the humerus is made using a very acceptable positioning with the dog in dorsal recumbency, but the antebrachium is at right angles with the cassette, making evaluation of the elbow difficult

Summary:

This fracture presents a challenge because of the large butterfly fragment. The lag screws achieve interfragmentary reduction and two additional screws provide anatomic reduction of the entire fracture. The K-E device is used for stabilization, with correct angulation of the fixation pins. Anatomic reduction and good stabilization of the fragments leads to a radiographic picture of fracture healing within 28 days.

However, the distal fixation pin was placed cranially within the condyle (white arrowhead) and caused destruction of the articular surface. On the lateral view of the elbow joint made after removal of the K-E device, secondary joint disease is evident.

So, a challenging fracture was treated in an appropriate mode with excellent healing, but the pin placement was not equal to the good planning, and the application failed because of the misplacement of a single fixation pin. It is not enough to work toward healing of the primary injury, the bone fracture; it also is important to ensure that no damage is done to the surrounding tissues. This dog will have permanent elbow lameness because of the arthrosis secondary to the fracture treatment.

Signalment: 2 years old, female, intact, Afghan Hound

History: dog was found recumbent, unwilling to bear weight on the right forelimb

Physical Examination: dog recumbent, crepitus in right forelimb

Radiographic Examination: right humerus

Study at time of entry (day 1):

1. badly comminuted segmental fracture of midshaft right humerus, with longitudinal splitting of the segmental fragment
2. several large butterfly fragments
3. shoulder and elbow joints appear radiographically normal

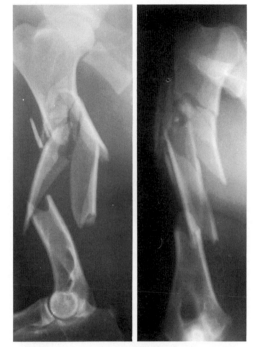

Postreduction study (day 30):

1. single interfragmentary cancellous screw
2. single cerclage wire placed proximally
3. large contoured bone plate placed caudolaterally
4. early callus formation proximally and distally at the site of the cancellous graft
5. midshaft fragments without callus formation (arrowheads)
6. excellent apposition and alignment of fragments

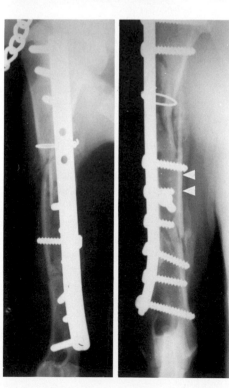

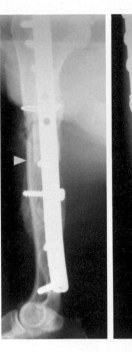 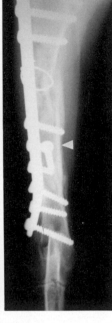

Postreduction study (day 50):

1. metallic devices remain as before
2. advanced callus formation
3. fracture lines still partially visible
4. persistent delay in participation by the fracture fragments in the healing process (arrowheads)
5. cerclage wire is noted to be buried caudally and medially by new bone
6. continued excellent apposition and alignment of fragments

Comments on radiographic findings:

1. pattern of callus formation seen in this patient is typical of that found with a high-energy trauma fracture—midshaft fragments are probably without a blood supply and rest in a severely traumatized soft tissue bed; thus, the ingrowth of the extraosseous capillary bed is slow and the bony response (callus) is delayed compared to the callus forming at the vascularized ends of the major humeral fragments
2. note the highly visible cortical shadow in the midshaft (arrowheads) on the last study; with subsequent healing of the fracture, this was proven to be only delayed vascularization of a cortical fragment, but could mistakenly have been assumed to be a sequestrum
3. a cancellous bone graft was used in this patient, probably confusing the evaluation of early callus formation

Summary:

This is a large dog with a bad fracture. Afghan Hounds appear to have bone that fragments easily and this, unfortunately, is an excellent example of it. Because of the character of the fracture, the postsurgical treatment may be appreciably longer and more difficult. The surgeon reconstructed the bone and then applied a neutralization plate. Lag screws and a single cerclage wire were used for the reconstruction. Some of the lag screws were placed through the plate. It might have been possible to consider lagging the bone fragments further proximally. With three screws proximally and three screws distally, the plate acts both as a buttress plate and a neutralization plate. Note that the plate was accurately contoured.

This is a monumental piece of work using the only treatment method available for a fracture of this type in a young Afghan Hound. It is recognized that healing is occurring in a delayed manner because of the high-energy trauma, presence of avascular fragments, and the extensive surgical trauma required to achieve reconstruction.

The patient continued to use the limb and at last report was exercising without lameness. This is a good example of what can be accomplished in an apparently hopeless case.

Signalment: 4 years old, female, spayed, Puli

History: right humeral fracture 25 days ago was treated by reduction and fixation using a single threaded-tip IM pin

Physical Examination: nonweight bearing on right thoracic limb; tip of IM pin can be palpated under skin of right shoulder

Radiographic Examination: right humerus

Study at time of entry (single lateral view) (day 25):

1. chronic midshaft fracture of humerus with one large butterfly fragment without solid bony healing
2. bony fragments appear to have impacted
3. single IM pin inserted proximally extends to just distal to the fracture site, and obviously has backed-out
4. shoulder and elbow joints are difficult to evaluate radiographically—new bone has formed proximally, presumably at the site of entrance of the IM pin (arrowheads); new bone response at the medial epicondyle is possibly posttraumatic in origin

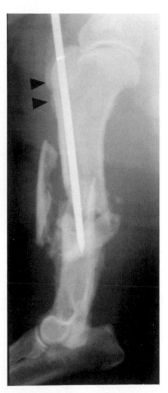

Study (single lateral view) (day 55):

1. IM pin has been removed
2. heavy callus forming around fracture site
3. butterfly fragment partially incorporated in healing callus
4. fracture line still identified—potential nonunion was considered
5. cranial angulation of distal fragment

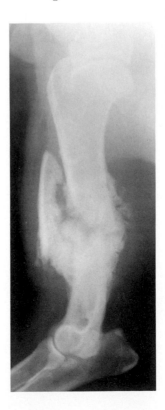

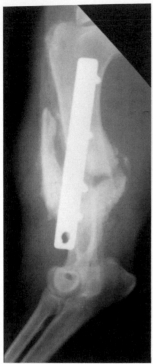 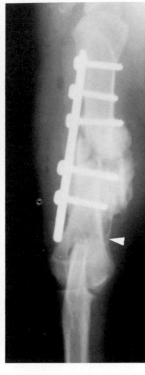

Postreduction study (day 59):

1. noncontoured bone plate with cancellous screws was used to stabilize a potentially nonunion fracture—three screws proximally, one screw securely in the distal fragment, and one screw near the fracture line
2. only 25% end-to-end apposition of fragments
3. cranial angulation of distal fragment
4. persistent bony response to IM pin placement in proximal tubercle
5. clinic records to not clarify purpose of drill hole in distal condyle
6. plate malpositioning prevented use of drill hole (arrowhead) for placement of the third screw in distal fragment
7. soft tissue gas secondary to surgical procedure

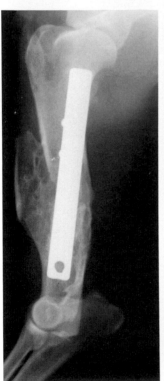 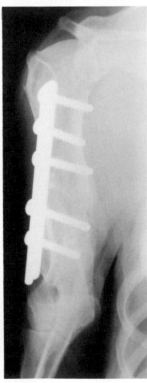

Postreduction study (day 285):

1. plate and screws remain as before
2. callus has filled area between fracture fragments and has incorporated butterfly fragment
3. cortical pattern or medullary cavity not yet reestablished

Comments on radiographic findings:

1. responsive callus formation seen on day 25 indicates motion at the fracture site during healing and does not indicate infection
2. at 55 days, the surgeon concluded that further callus development was unlikely and that there was a possibility of nonunion—hence, the fracture was plated
3. final exuberant callus formation shows healing in the presence of fragment motion
4. further modeling of the prominent bone fragment attached cranially will occur with resorption of the bone and redundant callus and subsequent restoration of the cortices

Summary:
The use of a small IM pin to repair a fracture of this type was incorrect. The location of the pin at the time of referral suggests that it had backed-out and was nonsupportive. Removal of the pin was performed in the hope that healing might progress in the face of an unstable fracture. It was quickly decided to achieve a more stable fixation using a bone plate. The plate was not contoured because of the exuberant callus formation. The cranial angulation of the distal fragment was not correctable because of the callus formation. Placement of a most distal screw was thus impossible. The plate was secured with three screws proximally. The longer fourth screw was placed through the fracture site and incorporated into the healing callus around the medial fragment. Only the single distal screw passed through both cortices. However, in spite of difficulties, the fracture proceeded to heal, with incorporation of the large, cranially located butterfly fragment.

The inept handling of the primary fracture led to a difficult situation in which nonunion was a probable outcome. Application of a bone plate proved that if stability is achieved, even a case such as this can have a successful outcome.

CASE 16

Signalment: 8 years old, female, intact, domestic short-haired cat

History: acute lameness in left thoracic limb after use of this bone as a site for collection of bone graft 1 day earlier

Physical Examination: nonweight bearing on left thoracic limb

Radiographic Examination: left humerus

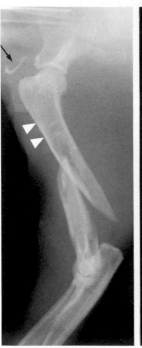
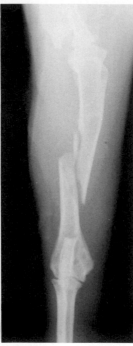

Study at time of entry (day 1):

1. spiraling fracture of midshaft humerus with two small cortical fragments
2. overriding of the major fragments with cranial angulation of distal fragment
3. both shoulder and elbow joints appear radiographically normal
4. note the normally ossified clavicle (arrow)

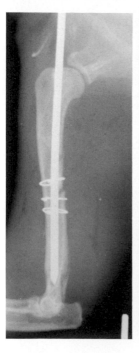
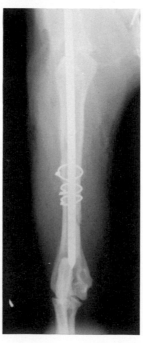

Intraoperative study (day 1):

1. single IM pin
2. three cerclage wires used to support the major fragments
3. the IM pin had not been cut on this intraoperative study
4. postsurgical soft tissue gas

Comments on radiographic findings:

1. the bone weakening after harvesting of a cancellous graft must have extended to midshaft based on the thinning of the cranial cortex (arrowheads)
2. note that the notch cut into the cortex to assist in preventing the cerclage wires from sliding cannot be seen

Summary:

This is an induced iatrogenic fracture after harvesting of a cancellous graft used in an arthrodesis of the tibiotarsal joint. Note how the perfectly straight humerus in the cat permits the IM pin to be inserted close to the humeral head proximally, run distally, and still remain within the center of the medullary cavity. It is possible in error to drive the pin into the elbow joint. In the dog, the IM pin tends to lodge in the medial epicondyle of the humerus. The diameter of the pin should be commensurate with the distal diameter of the shaft, which has the smaller measurement. Use of this guide prevents splitting of the bone. The pin is commonly removed after fracture healing.

The cerclage wires were used to bind the proximal and distal fragments. Wires should not be placed at the fracture site. Because the shaft is wider proximally, it is possible for wires to slip distally toward the thinner shaft. A K wire placed transversely, or, as in this patient, a notch cut in the bone, will prevent slipping. Cerclage wires have received bad publicity in the past because of their incorrect usage. If the wires are placed loosely, they move along the shaft of the bone, destroying the ingrowing capillary bed and leading to motion and subsequent nonunion. When the wires are placed tightly, the only vessels affected are those few directly beneath the wire, which does not interfere with bone healing. The use of cerclage wires is a valuable adjunct to fracture treatment.

The surgeon needs to be aware of the possibilities of creating a pathologic fracture at the site of removal of a cancellous graft. The site of graft collection appears to be quite large and too distal. Fortunately, this fracture was treated correctly and the patient made a complete recovery. The arthrodesis also was successful.

CASE 17 **Signalment:** unknown age, male, intact, domestic short-haired cat

History: cat was trapped beneath the hood of the car and beaten by the fan blades—brought to clinic by owner 3 days after the injury

Physical Examination: nonweight bearing on right thoracic limb, marked soft tissue swelling in region of elbow, soft tissue swelling near right antebrachiocarpal region and ventral thorax

Radiographic Examination: right elbow

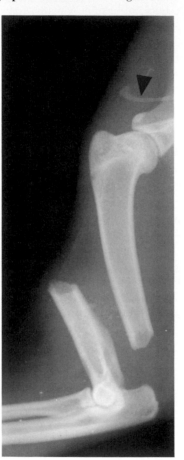

Study at time of entry (single lateral view) (day 1):

1. transverse mid-diaphyseal humeral fracture
2. marked overriding fracture fragments
3. soft tissue swelling at fracture site
4. shoulder and elbow joints appear radiographically normal
5. stage of physeal closure suggests a cat of 14 to 16 months of age

Postreduction study (day 3):

1. placement of full-length IM pin
2. type I K-E apparatus with one fixation pin proximally and one fixation pin ideally placed transversely in the distal condyle in a correct manner (1/1)
3. bayoneting of distal fragment onto the proximal fragment
4. good apposition and alignment of fragments
5. postsurgical soft tissue gas

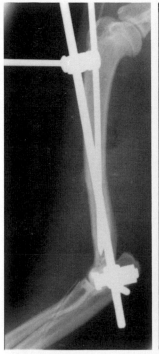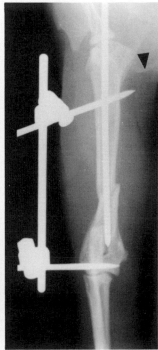

Postreduction study (day 70):

1. IM pin remains
2. removal of K-E apparatus
3. delayed callus bridging the fracture site with fracture line still visible
4. humerus 0.5 cm shorter
5. focal lucency in proximal humerus indicating pin track
6. pin track identified within the condyle

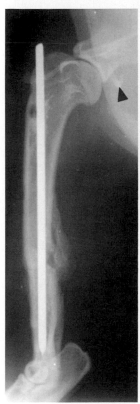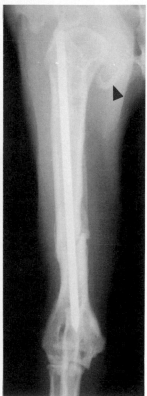

Comments on radiographic findings:

1. it would have been difficult to obtain the second view at the time of first radiographic examination—it is recommended to wait until surgery to obtain the second view
2. study the status of the joints proximal and distal to the injury both at the time of the trauma and after surgery
3. note the shift in position of the ossified clavicle and imagine how it can be confused for a foreign body or bony fragment (arrowheads)

Summary:

This is a rather typical fracture that looks simple and easy to reduce but is loaded with potential problems. The IM pin should have been larger to avoid the collapse of the fragments and lateral angulation at the fracture site. However, bayoneting of fragment ends is often an acceptable technique to obtain fragment stabilization, but may not have been necessary in this patient. Given the nature of the fracture, rotation at the fracture site is probable. The use of the external K-E device is a good choice to counter that rotation. The placement of the distal pin is especially good, considering that the spindle-shaped condyle in the cat demands that pin placement be perfect. The removal of the external fixator permits earlier ambulation of the patient and allows more stress stimuli to be applied to the bone, encouraging healing. This is referred to as "dynamism."

Usually, the proximal fixation pin is placed caudally within the heavier bone of the proximal humerus. It is now realized that the use of a single fixation pin proximally and distally and a single connecting bar (1/1) does not provide as great a degree of stabilization as one would expect. The use of two pins within each fragment in an active patient is advisable (2/2). The nature of callus formation and the collapse at the fracture site is indicative of motion and the probable cause of the delayed healing; still, the fracture healed successfully.

Radius and Ulna

Signalment: 3 months old, female, intact, mixed breed Collie

History: dog hit by car

Physical Examination: nonweight bearing right forelimb, crepitus in midshaft radius and ulna

Radiographic Examination: right antebrachium

Study at time of entry (day 1):

1. oblique, highly comminuted fracture at the junction of proximal and middle thirds of the radius with separation of the fracture fragments
2. large butterfly fragments from the radius (arrowheads)
3. long oblique fracture junction of middle and distal thirds of the ulna with separation of the fracture fragments
4. elbow and antebrachiocarpal joints appear radiographically normal
5. growth plates open, as expected for a puppy of this age

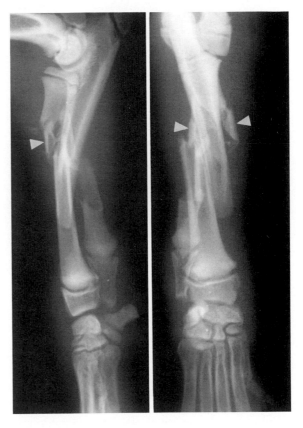

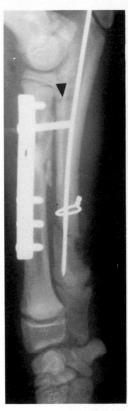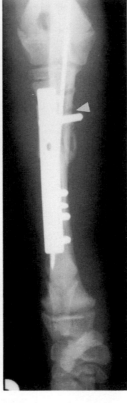

Postreduction study (day 2):

1. six-hole bone plate used to fix and stabilize radial fractures
2. two screws in proximal fragment
3. second screw angled to include the proximal third of the ulna
4. three screws engage the distal radius
5. IM pin placed within the ulna, driven normograde
6. cerclage wire reapproximates the proximal portion of the distal fragment that was found to be separated at surgery
7. cancellous graft at distal ulnar fracture site
8. note marked external rotation of paw

Comments on radiographic findings:

1. note the difficulty of determining the origin of the fracture fragments
2. physeal injury in a young puppy is difficult to estimate because of the thick, radiolucent growth cartilage—possibility of premature closure or delayed growth should be made known to the owner
3. no problem with single smooth pin crossing growth plate of proximal apophyseal center of the ulna
4. approximately 30° external rotation of the foot is noted on the postoperative study—best indicated by the radiographic appearance of the accessory carpal bone in relation to the humeral condyle surfaces
5. second screw markedly angled in an effort to enter the ulna, but scarcely engages the cortex—note the drill hole in the ulna adjacent to the screw (white arrowhead), and note the more proximal drill hole in the ulna at the level of the first screw (black arrowhead)
6. simple, oblique, distal metaphyseal fractures of second, third, and fourth metacarpal bones with minimal malalignment of fragments were noted on additional radiographs, and required no further treatment other than coaptation splinting

Summary:

This fracture is difficult because it is more proximal than usual with a resulting short proximal radial fragment; it is the result of high-energy trauma. The use of a bone plate is appropriate, however, the introduction of a long screw passing through the radius into the ulna invariably results in a synostosis. This is particularly inappropriate in a young growing animal. The screw may not be well seated, however. Distally, three screws enter the distal radial fragment. Having failed to obtain good security within the proximal radial fragment, the decision was made to place an IM pin normograde from the olecranon. The pin was small with a threaded tip so that it could easily enter the distal medullary cavity. The tip of the pin just caught the distal fragment. After cutting, the proximal tip of the pin appears long because of the thickness of the triceps tendon. The single cerclage wire was used to maintain position of an ulnar fragment detected at surgery.

In a dog of this age, the ulna is a large enough bone to deserve individual treatment. Unfortunately, there is almost 30° lateral rotation of the distal fragments after reduction. This can be avoided by ensuring that the foot is in correct anatomic position through observation of its relationship with the humerus at the time of surgery; observation of the fractured bone through the incision site does not always provide this information. The major technical problems are the placement of the proximal screw and outward rotation of the paw. The IM pin can be removed after healing.

The physis of the distal ulna is extremely delicate, and seemingly insignificant trauma can result in premature closure and subsequent deformity of the thoracic limb. This is considered a Salter-Harris type 5 injury.

CASE 2 **Signalment:** 8 months old, male, intact, mixed breed Cocker Spaniel–Poodle

History: dog fell from owner's arms

Physical Examination: marked pain elicited on palpation of right carpus–suspected sprain or carpal luxation

Radiographic Examination: right antebrachium

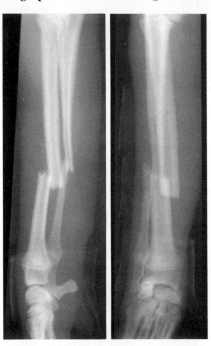

Study at time of entry (day 1):

1. low-energy transverse fractures of midshaft radius and ulna
2. overriding of fragments
3. minimal soft tissue swelling
4. elbow and antebrachiocarpal joints appear radiographically normal
5. open growth plates, as expected at this age in a small breed

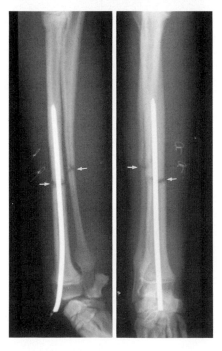

Postreduction study (day 2):

1. smooth IM pin was introduced into the medullary cavity of the distal fragment of the radius and is well seated in the proximal fragment
2. good apposition and alignment of the radial and ulnar fragments
3. resulting slight separation of radial and ulnar fracture fragments (arrows)
4. ulnar fracture not treated
5. metallic soft tissue sutures at fracture site and carpal region

Postreduction study (dog lame at this time) (day 50):

1. IM pin as before
2. heavy external callus formation around radial fracture site
3. good callus formation around ulnar fracture site
4. destruction of articular bone at radiocarpal, intercarpal, and carpometacarpal sites (suspected infectious arthritis and osteomyelitis)
5. soft tissue swelling in carpal region

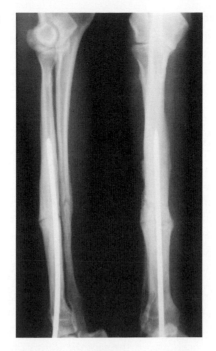

Postreduction study (day 90):

1. healed radial fracture with modeling of callus
2. IM pin remains as before
3. healed ulnar fracture with incomplete modeling of bone
4. marked soft tissue swelling in carpal region
5. substantial repair of subchondral bone destruction in carpal bones

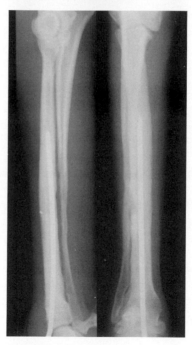

Comments on radiographic findings:

1. destructive changes in the carpal region were generalized and suggestive of infection; if they were posttraumatic secondary to pin placement, they should have been very localized and limited to the radiocarpal joint
2. osteopenia from disuse increased the likelihood of interpretations of osteomyelitis.
3. soft tissue swelling at the carpal area could have been secondary to the presence of the pin or could have been related to an infectious process
4. closure of the distal radial and ulnar physes is expected at this age

Summary:

This is an absolutely typical history, with a small puppy falling from owner's arms. The resulting fracture is very common in the midshaft to distal location. Delayed healing often is seen due to difficulty in maintaining fixation. The treatment choice was to introduce an IM pin through an entry hole in the distal portion of the distal radial fragment. The pin passed across the fracture site and was secured into the proximal fragment. Slight distraction was noted at the fracture site and probably was due to the forceful introduction of the pin.

It is strongly recommended that the distal tip of the pin be bent cranially and cut shorter than seen in this patient; the best technique is to create an acute angle to eliminate the possibility of damage to the radiocarpal joint. The tip of the pin could have been driven through the joint surface and countersunk to avoid the problem of joint injury.

A better choice of treatment would have been the application of a small bone plate, which would provide adequate stability, both axial and rotational control, and ensure early healing and return of the limb to function. In those cases in which plating is not feasible, fractures of this type lend themselves very well to external fixation splints.

At day 50, healing of the fractures is progressing, with fracture lines still visible. However, this constitutes delayed healing in a puppy with this type of low-energy fracture, where conditions lead to early healing with a large callus forming.

The changes noted radiographically affecting the carpal region are troublesome sequelae to the protruding tip of the IM pin. Although this is probably due to the repeated gouging of the pin on the cartilage and bone, the possibility of osteomyelitis and infectious arthritis needs to be considered. The soft tissue swelling could be expected in either situation. Activity of the dog plus failure to put the limb in a splint increased the degree of trauma to the opposing bone.

On the last study, the fracture healing is complete and callus modeling is present. Damage to the carpal joint still is present; however, the osteopenia that was present before is now gone. The IM pin must be withdrawn to allow for healing and to prevent further damage to the carpal joints. The limb should receive further protection from excessive activity through placement in a small Robert Jones splint. The soft tissue carpal swelling remains.

The best choice of treatment for the fracture would have been the use of a bone plate. If the fracture were more distal or the soft tissue injury more severe, external fixation should have been considered. An IM pin could be used as a third choice.

The fracture healed, but permanent damage was done to the joint.

Signalment: 2 years old, male, intact, Irish Setter

History: dog left home, returning the next day nonweight bearing on the right forelimb

Physical Examination: open fracture of the radius and ulna

Radiographic Examination: right antebrachium

Study at time of entry (day 1):

1. essentially transverse fractures at the junction of the middle and distal thirds of the radius and ulna
2. minimal comminution
3. medial angulation of the distal fragments
4. additional transverse fracture of distal ulnar metaphysis with minimal displacement of the fragments (arrow)
5. elbow and antebrachiocarpal joints appear radiographically normal

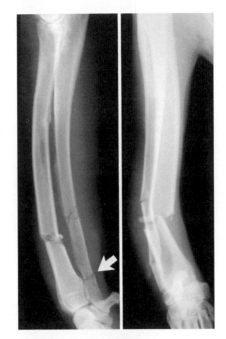

Postreduction study (single lateral view) (day 3):

1. bone plate placed cranially on radius (note drill hole [arrow] for jack used for dynamic compression)
2. good reduction of radial fragments
3. ulnar fractures left untreated with good fragment approximation

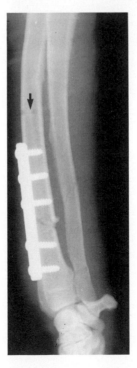

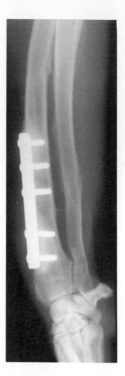

Postreduction study (single lateral view) (day 120):

1. radial plate and screws remain as before
2. radial fracture healed with no evidence of external callus formation
3. ulnar fractures healed with modeling of a minimal external callus
4. apposition and alignment of fragments remains good
5. distal ulnar fracture healed

Comments on radiographic findings:

1. the finding of the second ulnar fracture suggests a high-energy fracture and a greater degree of soft tissue injury
2. the more proximal drill hole in the radius (arrow) is associated with use of a separate device to obtain compression at the fracture site
3. the sclerotic bony response around the first, fourth, and fifth screws probably is associated with stress bearing shared between the plate and the bone; marked stress protection is present around the center of the plate, as indicated by the osteopenia
4. bone production surrounding a screw may be associated with chronic infection—the lack of any periosteal new bone tends to rule out this possibility
5. there is some radiographic evidence of minimal external rotation of the foot, which is evaluated more accurately clinically

Summary:
This fracture resulting from trauma in a large, young, active dog requires good stabilization for healing. The surgeon selected a suitably sized plate (4.5-mm screw hole). Screws should penetrate the far cortex by 2 mm. The apparent failure of the screws to penetrate the far cortex as seen on the radiographs is due to limb positioning. The x-ray beam must be perpendicular to the screws to reveal their true position. The screw holes occupy a considerable area of bone and tend to weaken it. If the screw holes are in a line, there is greater amount of weakening; this can be reduced by staggering the angles of the screw positions.

In this patient, the undisplaced ulnar fragments were well supported by the stabilized radius and required no additional fixation. In contrast, ulnar fractures in the cat usually are provided fixation.

In the absence of complications, the plate may not need to be removed. However, if pain should occur months to years later as a result of a loose screw, this would indicate the need for plate removal, and the limb would require support at this time. This can be provided by a Robert Jones splint. Often there is need to remove a portion of the screws at intervals to allow for strengthening of the bone, especially when it is not possible to control the patient's activity.

Signalment: 8 months old, female, intact, German Shepherd Dog

History: dog was hit by a car

Physical Examination: open wound on right forelimb with radial and ulnar fractures

Radiographic Examination: right antebrachium

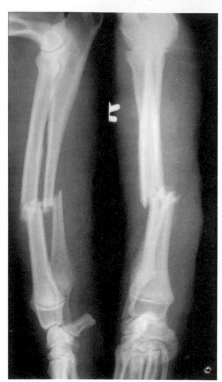

Study at time of entry (day 1):

1. slightly oblique midshaft fractures of radius and ulna
2. minimal comminution
3. moderate angulation with minimal contact of fragment ends
4. prominent soft tissue swelling
5. elbow and antebrachiocarpal joints appear radiographically normal
6. open growth plates, as appropriate for age in a large breed

Postreduction study (single lateral view) (day 12):

1. bone plate placed cranially on radius
2. note the first, fourth, and eighth screws penetrate through to the ulna
3. excellent reduction of radial fragments
4. radial fracture lines difficult to identify
5. note the dense cortical fragment (arrow)
6. heavier callus forming around un-treated ulna is related to periosteal stripping
7. ulnar fracture lines are difficult to identify
8. apposition and alignment of fragments are good
9. note suspected early closure of growth plates

Postreduction study (single lateral view) (day 34):

1. radial plate and screws remain as before
2. radial fracture lines less evident
3. persistent dense cortical fragment (arrow)
4. ulnar fracture with heavy maturing callus
5. ulnar fracture lines less evident
6. apposition and alignment of fragments remain good
7. growth plates are closed

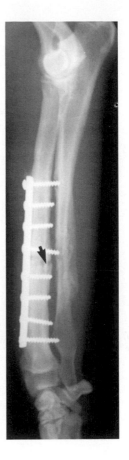

Postreduction study (single lateral view) (day 280):

1. recent removal of radial plate and screws, with screw holes evident
2. healed radial and ulnar fractures
3. lucency in midshaft of radius suggests stress protection by plate
4. cortical fragment has been incorporated
5. ulnar fracture healed with callus modeling
6. two areas of synostoses formed
7. distal ulnar osteopenia due to lack of weight bearing
8. apposition and alignment of fragments remain good
9. marked soft tissue atrophy

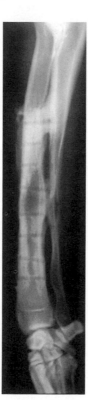

Comments on radiographic findings:

1. a fracture of this nature in the radius and ulna of a skeletally immature patient requires that the owner be warned of the possibility of premature closure or delayed physeal growth
2. change in appearance of the growth plates in 12 days may suggest a problem and should be monitored through radiographs of the opposite limb; at this age there is little remaining growth in length of the bones, and the clinical result of closure would be minimal
3. difference in amount of callus at the fracture sites illustrates the firm fixation of the radius and the minimal movement of the ulnar fragments during healing
4. stress protection is very evident in this patient, with marked osteopenia at the fracture site and distal ulna
5. it is suggested that the synostoses formed at the site of the longer penetrating screws

Summary:

The choice of the bone plate to treat this fracture was good. The length and size of the plate are satisfactory, with four screws proximally and four screws distally. The screws have been placed at varying angles so that the drill holes would not be in a straight line, which would have resulted in the creation of a weakened zone in the bone. However, three screws have penetrated through into the ulna, causing synostosis. Note the marked ostenpenia of the distal ulna and centrally at the radial fracture site.

The plate should be contoured to the slight bow of the radius so that there is no space between the plate and bone. In this patient, the contouring is difficult to identify. Removal of the plate shows the character of the cortical bone and suggests that the limb will need protection until strengthening occurs.

The fracture could have been treated with an external K-E device if the owner and surgeon were willing to accept a longer healing time. Treatment of that type would have required more care on the part of the owner and the clinician. Any presumed cost advantage of an external splint often disappears as a result of the extra treatment required.

Signalment: 3 years old, male, intact, Australian Shepherd Dog

History: dog hit by car and treated at another clinic—eventually presented for subsequent treatment 4 months later

Physical Examination: solid fracture callus palpated—no pain; limb with lateral angulation at the fracture site

Radiographic Examination: left antebrachium

Original study at referring clinic (single lateral view) (day 1):

1. oblique fractures at junction of proximal and middle thirds of the radius and ulna with minimal comminution and overriding of the fragments
2. minimal soft tissue swelling
3. elbow and antebrachiocarpal joints appear radiographically normal
4. limb supported by Mason metasplint

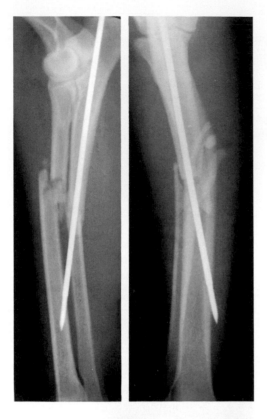

Postreduction study at referring clinic (day 2):

1. single, large, threaded IM pin introduced normograde into the proximal ulnar fragment but failed to enter the distal ulnar medullary cavity
2. limb angled laterally at fracture site
3. subcutaneous gas

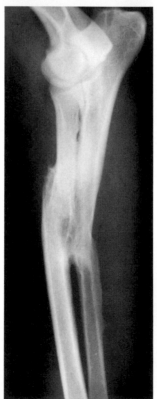

Postreduction study (single lateral view) (day 140):

1. IM pin removed
2. malunion healing with synostoses

Comments on radiographic findings:

1. failure to include the distal joint on the last radiograph study leaves us wondering about the degree of rotational deformity and angulation of the distal part of the limb

Summary:

The original repair probably did not use an open reduction. However, the tip of the pin should have been readily palpable and noted to be in an unacceptable position. The screw-tipped IM pin was introduced in a normograde direction down the ulnar shaft, missing the distal portion medially and ultimately inserting near the periosteum of the distal radial cortex. It should be noted that insertion within the fibrous tissues may give the surgeon a feeling of stability. Having observed the misplacement of the pin on the immediate postoperative radiographs, one questions why the surgeon did not correct the error.

The final study 140 days later was to evaluate the deformed limb for corrective orthopedic surgery. Rectification would be difficult, or almost impossible, because the massive fusion between the radius and ulna would make identification and manipulation of the component parts difficult. In this patient, correction of angulation and rotational deformity would require a great deal of expertise.

This typical radial and ulnar fracture offers difficulties because of the great tendency for the distal fragment to rotate; therefore, it requires firm fixation. Ideally, this large dog would have been a good candidate for bone plating. Surgical exposure would have provided for adequate reduction and the placement of a plate for maximal control. The use of a single ulnar pin, even if placed correctly, would have had potential for failure because of persistent rotational motion at the radial fracture site.

Signalment: 3 years old, male, castrated, domestic short-haired cat

History: cat found by side of road, presumably hit by car

Physical Examination: cat depressed and unable to respond to stimuli, crepitus noted in right forelimb

Radiographic Examination: right antebrachium

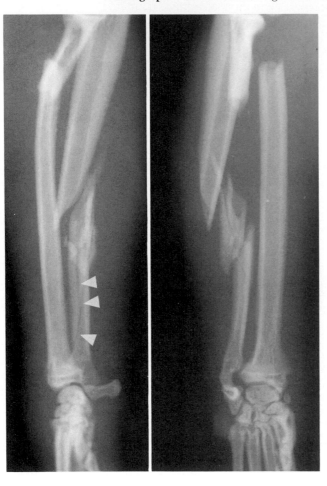

Study at time of entry (day 1):

1. badly comminuted fracture of middle and distal thirds of ulna with large, longitudinally split butterfly fragment
2. no apposition of ulnar fracture fragments
3. transverse fracture in proximal third of radius
4. overriding of radial fracture fragments
5. extensive soft tissue swelling
6. elbow and antebrachiocarpal joints appear radiographically normal

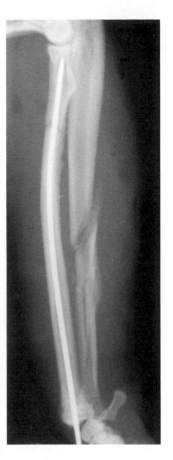

Postreduction study (single lateral view) (day 2):

1. single threaded IM pin positioned in a retrograde manner in radius (pin not cut)
2. good reduction of radial fragments
3. ulnar fragments come into better position with limb lengthening
4. IM pin within the antebrachiocarpal joint
5. postsurgical soft tissue gas

Postreduction study (single lateral view) (day 40):

1. IM pin fractured at site of radial fracture (arrowhead)
2. impaction (3 mm) at radial fracture site with resulting small fracture fragments; proximal shifting of threaded tip
3. radial fracture line still visible
4. caudal angulation of distal radial fragment
5. tip of IM pin protrudes into the radiocarpal joint space, causing severe destructive arthrosis
6. early healing of ulnar fragments with fracture lines still visible
7. elbow joint remains radiographically normal

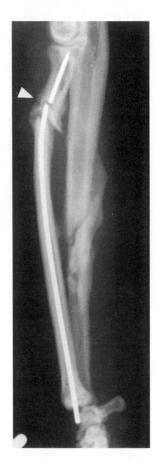

Comments on radiographic findings:

1. it is important to notice that the ulnar fracture lines extend distally into the metaphyseal region (arrowheads)
2. ulnar healing is a good example of how a fracture can attempt to heal without fixation or stability
3. the extent of the collapse of the fracture fragments can be ascertained radiographically by comparing the length of the radius with that of the opposite limb
4. the strange pattern of ulnar callus probably is influenced by periosteal injury

Summary:

These proximal radial and midshaft ulnar fractures are somewhat different from those usually seen, and are more difficult to treat because of the short proximal radial fragment and the large split segmental fragment in the distal ulna. Note the much larger size of the ulna in the cat than in the dog. It is more supportive of weight and usually requires separate treatment. The surgeon elected to treat only the radius and to introduce the IM pin through the radiocarpal joint and pass it into the proximal radial fragment. The radial bowing suggests a maximum tension on the pin. The end should have been countersunk to avoid disaster to the radiocarpal joint. The ulnar fracture was ignored.

Collapse of the ulnar fracture caused further tension on the radial pin, and the pin broke as a result of cycling. The injury to the radiocarpal joint is severe. At the time of the second study, the radial pin should have been withdrawn to prevent further injury. If the limb were then placed in a Robert Jones splint, the fractures probably would have healed in malunion.

The primary treatment might have used a toy plate on the ulna, even though the distal fragment was split. A small plate might have been used on the proximal radius, although exposure of this region is difficult and the radial nerve needs to be protected. Another approach might be the use of an external fixator using small fixation pins and two connecting bars (type II). This could be attempted even though the proximal fragment is small. It also might have been possible to use an IM pin in the ulna with cerclage wires to stabilize the fragments, in conjunction with a short IM pin in the radius introduced at the midshaft and passed proximally across the fracture line.

This is an example of the ill-advised use of a small IM pin for treatment of a proximal radial fracture, along with failure to treat the ulna, which placed all of the stress on the radius. Choice of an initial therapy that is not satisfactory may yield a result that is impossible to correct by subsequent surgical repair.

Signalment: 1 year old, male, intact, Labrador Retriever

History: dog was hit by a large mowing machine on a ranch

Physical Examination: limping on left forelimb—bones unstable with crepitus noted

Radiographic Examination: left antebrachium

Study at time of entry (day 1):

1. limb in a Mason metasplint with reasonable fracture reduction and alignment
2. short oblique fracture of midshaft left radius
3. transverse fracture junction of middle and distal thirds of the left ulna
4. 25% end-to-end apposition of the radial fragments
5. 100% end-to-end apposition of the ulnar fragments
6. elbow and antebrachiocarpal joints radiographically normal

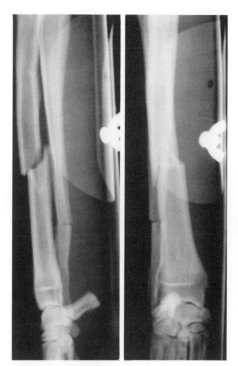

Postreduction study (single lateral view) (day 2):

1. single large IM pin with threaded tip positioned in a retrograde manner within the radius
2. new butterfly fragment at midshaft radius at site of original simple fracture (arrow)
3. good apposition and alignment of radial fragments
4. end of IM pin within the radiocarpal joint
5. distraction of ulnar fragments
6. external rotation of foot

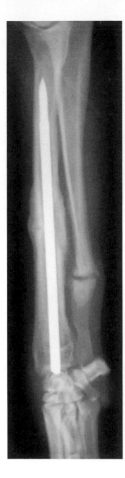

Postreduction study (single lateral view) (day 30):

1. IM pin remains in position
2. good external callus formation bridges radial fracture site and incorporates the butterfly fragment
3. radial fracture lines still visible
4. external callus forming around ulnar fragments
5. ulnar fracture line remains visible and widened
6. distal end of the pin remains in the radiocarpal joint space
7. external rotation of foot as before

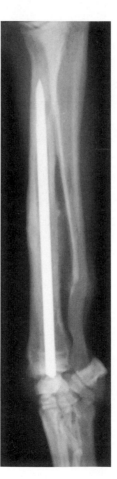

Postreduction study (single lateral view) (day 120):

1. IM pin remains in position
2. callus modelling resulted in bridging of the radial fracture site with incorporation of the butterfly fragment with closure of the fracture line
3. callus bridges the ulnar fracture gap
4. ulnar fracture line almost closed
5. distal end of the pin remains in the radiocarpal joint space
6. reactive bone in distal radius and radial carpal bone around pin end
7. focal soft tissue swelling at carpus
8. external rotation of foot as before

Comments on radiographic findings:

1. fracture may heal slowly due to the trauma-related loss of soft tissue external blood supply, and the large IM pin that destroyed the medullary blood supply
2. note healing of ulna without fixation, even with the gap
3. it is important to note the developing arthrosis
4. external rotation of foot in a large, active dog is of great concern

Summary:
The location of the radial fracture is common. It is important, however, to note that the fracture in the ulna is more distal, probably where the bone is weaker. It is important to question why the surgeon chose to use a screw-tipped IM pin and then, after cutting the pin short, tried to countersink it. Driving the pin proximally to position the end beneath the articular surface removes the value of the thread. In the process, a short end of the pin was left extending into the radiocarpal joint, where it gouged the opposite articular surface with movement of the foot. With the attempted impaction of the pin, the fracture line was widened. With the resulting impaction of fragments occurring later, the pin end protruded into the joint, resulting in damage to the radiocarpal joint. Perhaps the trauma of trying to countersink the end of the pin also resulted in the creation of the butterfly fragment at the fracture site that was not present earlier.

This is an example of a fracture that healed, even if in a delayed manner. The subsequent arthrosis within the radiocarpal joint must be taken into account. This is a large hunting dog that needed use of this limb for at least 10 more years. It is to be feared that the secondary posttraumatic joint disease will appreciably decrease joint function.

Removal of the screw-tipped IM pin will require cortical removal from the distal radius cranially to create a window that exposes the pin so that it may be grasped with dental forceps. This will permit the surgeon to rotate the pin slowly to unthread the tip. Entry into the radiocarpal joint then will permit use of needle-nosed pliers to grasp the cut end of the pin and retract it. Until this is done, the degree of secondary arthrosis will progress.

CASE 8

Signalment: 7 years old, female, spayed, mixed breed dog

History: dog fell 10 feet onto rocks while walking on a pipe over a ditch

Physical Examination: nonweight bearing on left forelimb

Radiographic Examination: left antebrachium

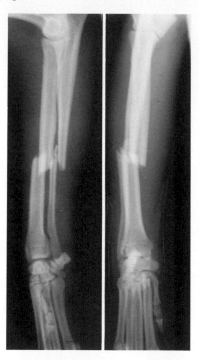

Study at time of entry (day 1):

1. near-transverse fractures of midshaft radius and ulna
2. overriding of fragments
3. minimal soft tissue swelling
4. elbow and antebrachiocarpal joints appear radiographically normal

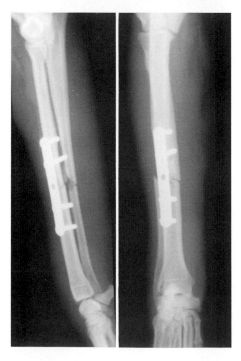

Postreduction study (day 1):

1. five-hole finger (mini) bone plate used to fix and stabilize radial fracture
2. two screws in proximal fragment and two screws in distal fragment
3. good apposition and alignment of fragments
4. butterfly fragment removed from ulnar fracture site

Postreduction study (day 12):

1. fractured radial bone plate
2. failure of callus development
3. malalignment of fracture fragments

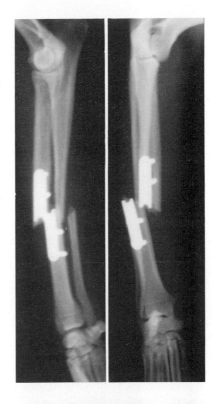

Second postreduction study (day 15):

1. removal of fractured plate
2. restabilization of radial fracture using a larger plate with three screws proximally and two screws distally
3. good alignment of fracture fragments

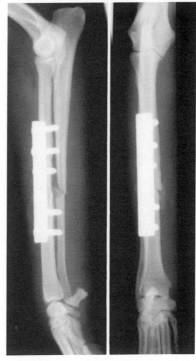

Comments on radiographic findings:

1. absence of callus on second postreduction study suggests that the plate had been broken for "several" days, with motion at the fracture site that discouraged any callus formation

Summary:

This is a typical fracture with an atypical history, with minimal soft tissue damage as indicated by the absence of swelling. Selection of a bone plate was excellent; however, the plate was too small and it fractured within several days. The fracture occurred through the open hole, which is the weakest part of the plate. Micromotion results in cycling and crystallization of the metal, which leads to subsequent fracture. There was no inherent error in plate manufacture or positioning. Although the screws held in this patient, it is advisable to place three screws proximally and three screws distally.

The next repair resulted in removal of the broken plate fragments, and a suitably sized plate was applied. The original screw holes were tapped larger to accept the larger screws. This avoided further weakening of the bone, and was possible because the distance between the holes in the two plates was similar. A longer plate with provision for an additional screw distally would have been ideal.

The use of a cancellous graft is highly recommended, almost mandatory, in treatment of radial and ulnar fractures in small dogs and cats. Note that no graft was used with the first plate application. Ultimately, the fracture healed and the plate was removed, with protection of the limb provided for 2 weeks after plate removal. This type of transverse fracture lends itself to treatment by a dynamic compression plate. Note that this type of repair spares the joint surfaces, and osteopenia is avoided because the patient is partially ambulatory during healing.

Signalment: 3 years old, female, intact, German Shepherd Dog

History: dog referred to the clinic with limb in a coaptation splint—injury occurred 2 days previously

Physical Examination: fractured right radius and ulna

Radiographic Examination: right antebrachium

Study at time of entry (day 2):

1. badly comminuted midshaft fractures of radius and ulna
2. several butterfly fragments
3. elbow and antebrachiocarpal joints appear radiographically normal

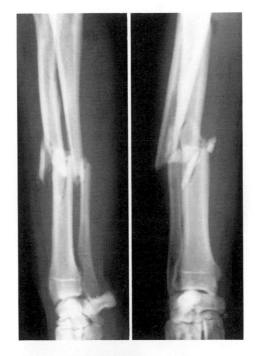

Postreduction study (day 2):

1. contoured bone plate placed cranially on radius
2. two interfragmentary screws
3. reduction of radial fragments with excellent apposition and alignment
4. ulna left untreated with gap at fracture site
5. cancellous graft resembles early callus formation
6. note missing bone fragments at center of fracture site (arrowheads)

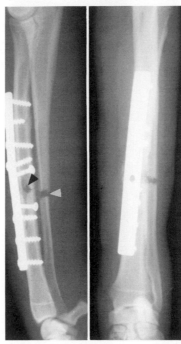

Postreduction study (single lateral view) (day 26):

1. radial plate and screws remain as before
2. radial fracture with external callus formation is beginning
3. radial fracture lines begin to disappear
4. ulnar fracture with heavy bridging callus formation
5. apposition and alignment of all fragments remain excellent
6. note the callus beginning to creep around the bone plate (arrowhead)

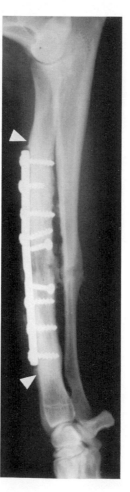

Postreduction study (single lateral view) (day 90):

1. radial plate and screws remain as before
2. callus formation begins to mature
3. callus forming around plate proximally and distally (arrowheads)
4. fracture lines not visible
5. ulnar fracture healing with callus modeling
6. synostosis is beginning to form
7. apposition and alignment of fragments remain excellent

Postreduction study (single lateral view) (day 230):

1. radial plate and screws remain as before
2. almost complete modeling of radial callus with restoration of medullary cavity
3. osteopenia at fracture site due to stress protection by the plate
4. ulnar fracture healed with modeling of ulnar callus
5. synostosis is complete
6. apposition and alignment of fragments remain excellent

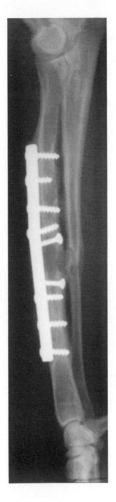

Comments on radiographic findings:

1. length of the fracture lines must be carefully studied in a case like this; obliquity of the fracture lines plus comminution plus the possibility of linear fracture lines markedly increase the size of plate required
2. loss of the bony fragment(s) creates a gap at the fracture site that requires a longer time to bridge
3. note that with strong fracture stabilization the patient uses the limb, and disuse osteopenia does not occur in the distal bones
4. note dense bone forming around the radial fracture site and how this becomes less dense on the last study—there probably is stress protection of the bone by the plate, so the radius does not require as much bone tissue as it normally would
5. note osteopenia in the distal ulna, probably secondary to the synostosis removing stress to that part of the bone
6. it is interesting to speculate why the synostosis formed, probably because the fractures were adjacent to each other

Summary:
On this big, active dog, the surgeon wisely elected to use a large bone plate. The smaller butterfly fragments were reduced and secured by lag screws. This is a good example of interfragmentary compression. Further bending of the plate so that the concave surface was placed next to the bone would have prevented the opening of the ulnar fracture that occurred in this patient. Through effective contouring, it would have forced the ulnar fragments together.

The increase in intramedullary bone in association with fracture healing seems to occur more in this breed of dog. Experience suggests that the bone in German Shepherd Dogs is more like "pine" wood, whereas other dogs have bone more similar to "oak" wood. The fracture healed nicely, and the osteopenia caused by the synostosis will be of little clinical concern.

An external K-E apparatus used alone could have been used effectively in a case of this type, if it were thought that the activity of the patient could have been controlled.

Signalment: 1 year old, female, intact, German Shepherd Dog

History: owner found lame dog

Physical Examination: left forelimb severely swollen from elbow distally; dog was febrile

Radiographic Examination: left antebrachium

Study at time of entry (day 1):

1. slightly oblique fracture at junction of middle and distal thirds of the radius
2. oblique fracture of midshaft ulna
3. marked overriding of fragments
4. minimal soft tissue emphysema (arrow)
5. elbow and antebrachiocarpal joints appear radiographically normal

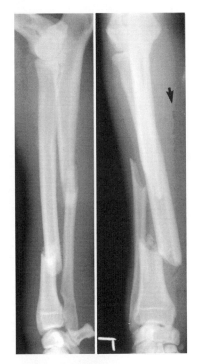

Postreduction study (single lateral view) (day 23):

1. bone plate placed cranially on radius with good reduction of the fragments
2. minimal bony response around the radial fracture site
3. radial fracture line still visible
4. single small pin placed in the ulna with good reduction of the fragments
5. pin enters the elbow joint
6. callus formation around the ulnar fracture
7. ulnar fracture line still visible
8. apposition and alignment of fragments are excellent

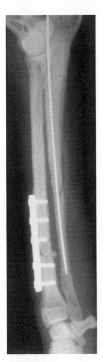

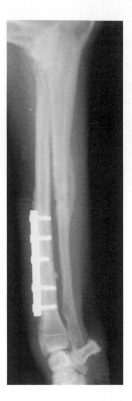

Postreduction study (single lateral view) (day 53):

1. radial plate and screws as before
2. radial fracture lines not seen
3. ulnar callus maturing
4. pin withdrawn from ulna
5. ulnar fracture line not seen
6. apposition and alignment of fragments remain excellent

Comments on radiographic findings:

1. the "bubbly" pattern of the soft tissue gas on the first study suggests muscle injury, whereas a linear pattern would suggest gas that has dissected within the fascial planes
2. the ulnar pin entered the elbow joint space; this should have been determined on an immediate postreduction radiograph and should not have remained for 3 weeks before being withdrawn
3. compare the absence of callus forming around the radial fracture, which has rigid stabilization, with the more exuberant callus forming around the ulnar fracture, which has only minimal stabilization
4. the callus around the ulnar fracture is referred to as a "bucket handle" callus and forms subperiosteally in accordance with the manner of stripping of the periosteum

Summary:

The choice of a bone plate is good, considering the obliquity of the fracture, with three screws placed proximally and two distally. The choice of an IM pin for additional stability was reasonable, but because it was positioned retrograde, it lacked a "guide" and entered the joint space. If it had been introduced proximally, the pin would have followed the medullary cavity. The effect of the pin within the joint even for a short period is evidenced by reactive bone formation, primarily noted periarticularly on the humeral condyle and radial head.

Note the bony response within the medullary cavity surrounding the pin on the first postreduction film. The pin was retracted as soon as its intra-articular location was noted, which unfortunately was 3 weeks after placement.

The fractures healed nicely with an absence of a disuse osteopenia. This indicates that the patient was able to use the limb during fracture repair.

Signalment: 5 years old, female, spayed, Miniature Poodle

History: dog was stepped on by owner

Physical Examination: marked external rotation of the foot with pain and crepitus

Radiographic Examination: left antebrachium

Study at time of entry (day 1):

1. slightly oblique fractures within the distal third of the radius and ulna
2. apposition of the fracture fragments ends
3. soft tissue swelling laterally
4. elbow and antebrachiocarpal joints appear radiographically normal
5. bandage holding splint in position

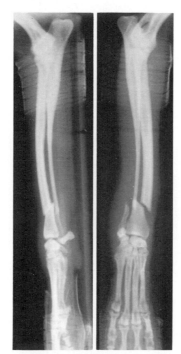

Postreduction study in plaster cast (day 5):

1. no end-to-end apposition of fracture fragments

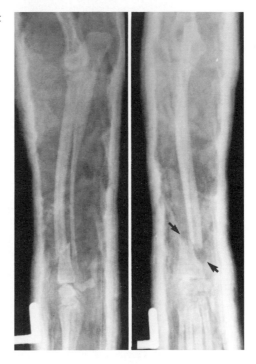

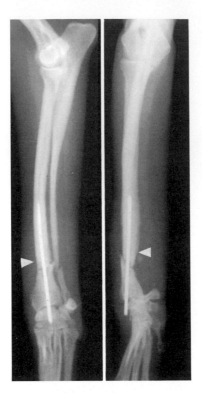

Postreduction study (day 30):

1. a single IM pin has been placed in the distal radius
2. no callus formation at radial fracture site
3. sclerotic appearance of distal end of proximal radial fragment (arrowheads)
4. IM pin protrudes into radiocarpal joint
5. minimal fragmentation at the radial fracture site
6. no callus formation around the ulnar fracture
7. 90° rotation of the distal fragment
8. disuse osteopenia of metacarpal bones

Comments on radiographic findings:

1. it is important to use both views in the determination of apposition and alignment of fracture fragments—note that on the study after use of a plaster cast, on the lateral view the ends of the fragments appear to be in apposition, whereas on the craniocaudal view there is almost no contact between the ends of the fracture fragments (arrows)
2. callus formation should not be evaluated on a radiograph made through a plaster cast; such a radiograph is suitable only for evaluation of the apposition and alignment of the fragments
3. the bony response within the proximal radial fragment is frightening when compared to the lack of response from the distal fragment—this is a nonunion ready to happen
4. injury to the radiocarpal joint is severe

Summary:

This is a very typical distal fracture of the radius and ulna that frequently occurs in small breeds with a history of low-energy injury. The 5- to 6-day delay in fixation with the limb in a cast is not acceptable. There is evidence that a cast does not provide sufficient support for a fracture of this type, and the radiograph shows the difficulty in maintaining end-to-end apposition of the fracture fragments. A radiograph should be taken immediately after fracture reduction to learn the status of the repair. In this patient, there probably never was good positioning of the fragments, and there was considerable delay in making the correction.

The surgeon elected to introduce a small-diameter IM pin through the dorsocentral portion of the distal radial fragment. One must remember that the cross section of the bone at this level is ovoid and cannot accept a larger pin. The pin was cut off but the end was not bent away from the bone. The pin probably would not pass further proximally because of the cranial bowing of the radius. The fracture needs an external splint to support it in addition to the pin. A cancellous graft would have been of value in a fracture of this type.

It is necessary to palpate the foot during reduction so that the dorsum of the foot points in the correct direction as compared with the elbow joint. In this patient, the external rotation of the distal limb is almost 90°.

The best treatment would have been to use the smallest plate with three screws proximally and two screws distally. An IM pin, even with a cancellous graft, would not have provided a very good chance of healing. Even external fixation, although recommended, because of the size of the bone and the need of extremely small pins, would not have resulted in as good healing as with use of a plate. The K-E device is not provided with a small enough pin for a dog of this size. It is possible to use straight, sharp cutting-edge needles for the pins and methylmethacrylate for the connecting bar in a type II formation.

I expect that this fracture will proceed to nonunion with resorption of the distal fragments. The brief window of opportunity before irreparable damage occurs dictates the immediate placement of a small bone plate with a cancellous graft as the best mode of treatment.

CASE 12 **Signalment:** 8 months old, female, intact, Brittany

History: dog caught her left forefoot in a door

Physical Examination: marked external rotation and lateral angulation of the foot indicating fracture or luxation

Radiographic Examination: left antebrachium

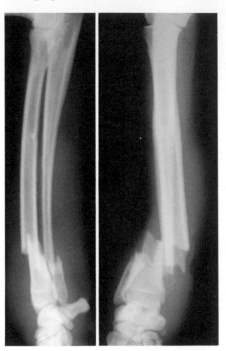

Study at time of entry (day 1):

1. slightly oblique fracture at the junction of the middle and distal thirds of the radius
2. transverse fracture within the distal ulnar metaphysis
3. overriding of fracture fragments
4. marked soft tissue swelling distally
5. elbow and antebrachiocarpal joints appear radiographically normal
6. growth plates are closing, as would be typical for a dog of this age

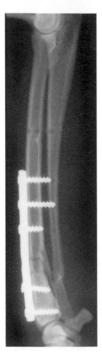

Postreduction study (single lateral view) (day 5):

1. plate placed cranially on radius
2. good reduction of radial fragments
3. note the previously unobserved radial fragment originating from the caudal cortex
4. ulnar fracture left untreated
5. good apposition and alignment of all fragments

Postreduction study (single lateral view) (day 38):

1. radial plate and screws remain as before
2. radial fracture healed
3. pencilling of the distal tip of the proximal ulnar fragment
4. proximal ulnar fragment uniting with radius
5. apposition and alignment of fragments remain good

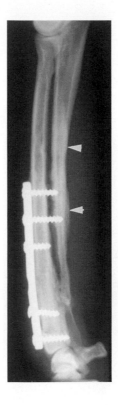

Comments on radiographic findings:

1. note two empty drill holes made at the fracture site that were wisely left unused
2. the most proximal drill hole was made for a separate device (compression jack) used to apply compression to the fracture fragments
3. first and second screws engaged the ulna
4. the sclerotic pattern in the midshaft of the ulna is associated with the healing of the proximal drill holes and the presence of screws entering the ulna—this pattern, along with the osteopenia of the distal ulnar fragment, represents the new pattern of stress lines that passes from the proximal ulna into the distal radius

Summary:

It is very common for fractures to have a "step" in the end that is convenient for obtaining anatomic reduction. The selection of a bone plate was good for a fracture of this type because the fragments fit with opposing surfaces and compression was possible. The size of plate selected was appropriate, with three screws proximally and two screws distally. The surgeon drilled two additional holes at the fracture site, but, prudently, the screws were not inserted. It was better that they were not placed. The position of the plate was good, without encroachment on the radiocarpal joint distally. The first and second screws were long, engaging the ulna, and appear to be leading to development of synostoses. The depth gauge must be used carefully because it is physiologically better to use screws of proper length.

   A possibility exists for a second, more distal site of synostosis centering around the empty drill holes and the impacted cortical fragments.

CASE 13

Signalment: 8 years old, female, intact, German Shorthaired Pointer

History: dog was hit by a car

Physical Examination: limping on left forelimb, skin laceration on distal antebrachium, marked soft tissue swelling

Radiographic Examination: left antebrachium

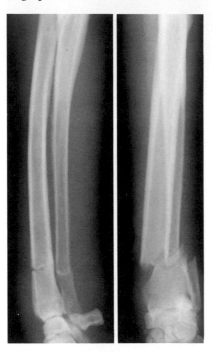

Study at time of entry (day 1):

1. transverse fractures of distal metaphyses of left radius and ulna with minimal comminution
2. 50% end-to-end apposition of the radial fragments
3. no end-to-end apposition of the ulnar fragments
4. marked soft tissue swelling
5. soft tissue laceration of medial aspect of limb
6. elbow and antebrachiocarpal joints radiographically normal
7. carpometacarpal joints with periarticular new bone lipping, suggesting minimal arthrosis of a chronic nature
8. suggestion of osteopenia at fracture site

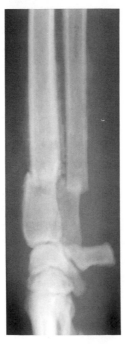

Postreduction study (treated with external cast to allow for soft tissue healing) (single lateral view) (day 20):

1. marked periosteal new bone (callus) on distal radius and ulna
2. collapse at fracture site
3. apposition and alignment of fragments essentially as before
4. persistent soft tissue swelling

Postreduction study (treated with bone plate) (single lateral view) (day 26):

1. six-hole bone plate with six fully threaded cortical screws bridging the distal radial fracture, with cancellous graft cranially
2. screw number four is at the fracture site (arrowhead)
3. fracture lines still present
4. greater amount of periosteal new bone (callus) formation on radius
5. periosteal response on proximal ulnar fragment with absence of bony response at the ulnar fracture site
6. distal end of the plate is at the radiocarpal joint space
7. disuse osteopenia of carpal bones
8. persistent soft tissue swelling
9. minimal external rotation of foot

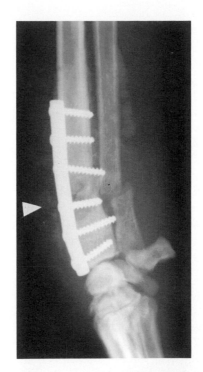

Postreduction study (single lateral view) (day 60):

1. bone plate bridging the distal radial fracture remains as before
2. lucency noted around screw number four at the fracture site
3. radial fracture lines still evident
4. periosteal new bone response on the radius as before
5. callus bridges distal ulnar fracture
6. ulnar fracture line still evident
7. distal end of the plate at the radiocarpal joint space causing marked destruction to the radiocarpal bone
8. persistent disuse osteopenia of carpal bones
9. soft tissue swelling less obvious
10. minimal external rotation of foot

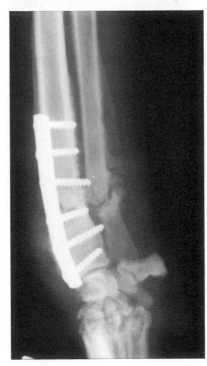

Comments on radiographic findings:

1. osteopenia in this older dog plus the nature of the fracture may suggest that the fracture is pathologic—later studies show that this is not a problem
2. because the early periosteal response (20 days postinjury) was so proliferative and widespread, the possibility of infection had to be considered; follow-up studies continue to suggest the possibility of osteomyelitis
3. last study still suggests osteomyelitis and suppurative arthritis
4. studies made 6 and 20 weeks later showed both fractures to be healed, without evidence of bone or joint infection
5. disuse osteopenia is evident in the distal carpal and proximal metacarpal bones
6. patient was lame on right elbow next year with ununited anconeal process—the right elbow was not radiographed at the time of trauma to the left radius and ulna—a previously clinically silent lesion may be exacerbated by additional stress as a result of favoring the opposite limb

Summary:

This is not an uncommon fracture in the forelimb of the dog. It is important to strive for firm stabilization of the fragments. It will not suffice simply to splint the limb because invariably the foot drifts laterally into a valgus deformation, with delayed healing. In this patient, it was decided to wait for 20 days before fracture fixation to permit soft tissue healing. This delay resulted in early callus formation, making it difficult for the surgeon to identify bony fragments and move them. The adherence of soft tissues to the developing callus requires that they be stripped away at the time of the delayed surgery. This destroys the developing extraosseous blood supply to the fracture, and the result is a delayed healing. Perhaps treatment should have been performed earlier.

The election to plate the fracture was reasonable, but the plate was positioned far too distally, with the tip overhanging and interfering with the radiocarpal bone. This happened through the attempt to position as many screws within the distal bony fragment as possible. In this patient, the last two screws were well placed in the distal fragment. It might have been better to use a longer plate and have four screws within the proximal fragment. Placement of the fourth screw at the fracture site is questionable, and the result cannot be good. Any motion destroys bridging callus and delays healing. It often is permissible not to treat ulnar fractures in larger dogs, where there is good fragment apposition, as was done in this case.

Placement of the plate was costly for this patient, with development of severe secondary joint disease within a short time. The resulting minimal external rotation of the foot is probably not of great concern.

It might have been a consideration in a radial fracture with a short distal fragment to use a type II external fixator with one threaded pin distally and three threaded pins proximally. This would permit applying some degree of compression to assist in fracture healing.

Signalment: 8 years old, male, castrated, mixed breed Golden Retriever

History: dog referred to the clinic with limb in a coaptation splint; limb had been in a splint since the injury 1 month earlier.

Physical Examination: chronic fracture of left radius and ulna

Radiographic Examination: left antebrachium

Study at time of entry (day 32):

1. chronic oblique fractures in distal thirds of the radius and ulna
2. butterfly fragments
3. minimal periosteal new bone
4. fracture edges are not sharply identified
5. no apposition of the fragments
6. elbow and antebrachiocarpal joints appear radiographically normal
7. minimal disuse osteopenia in distal limb
8. widening of radiocarpal and intercarpal joint spaces

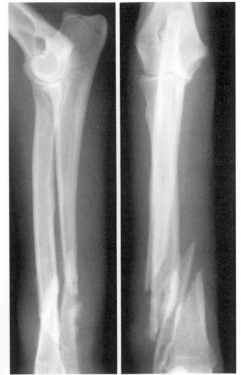

Postreduction study (single lateral view) (day 34):

1. plate placed cranially on radius
2. single interfragmentary screw
3. good reduction of radial fragments
4. dense, avascular cortical fragment within radial fracture site (arrow)
5. ulna left untreated with gap at fracture site
6. early callus formation on ulna fragments proximally and distally

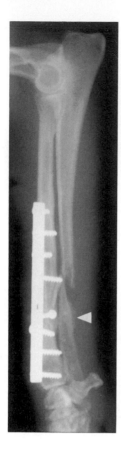

Postreduction study (single lateral view) (day 57):

1. radial plate and screws remain as before
2. minimal callus formation
3. dense, avascular cortical fragment persistent within radial fracture site
4. ulnar fracture fragments remain as before
5. minimal callus formation (arrowhead)
6. apposition and alignment of fragments remain good
7. disuse osteopenia is advanced

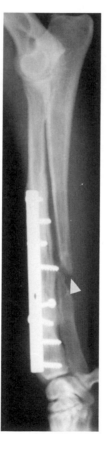

Postreduction study (single lateral view) (day 200):

1. radial plate and screws remain as before
2. radial healing with incorporation of cortical fragment
3. sclerosis around proximal and distal screws suggests weight bearing through these areas with stress protection at the fracture site—note synostosis at level of fifth screw
4. nonunion ulnar fracture with pencilling of fragment ends (arrowhead) and maturing of callus
5. apposition and alignment of fragments remain good

Comments on radiographic findings:

1. it is important to closely evaluate the character of the fracture fragments—in this patient, on the first study the fracture surfaces are indistinct and there is minimal new bone response on the larger fragments; the decision that this is a chronic fracture with early callus formation is supported by the clinical history and suggests the need for more aggressive treatment
2. osteopenia on the first study accompanies soft tissue atrophy due to disuse
3. the dense cortical fragment in the center of the radial fracture site is not a sequestrum but only an avascular cortical fragment that is delayed in joining in the fracture healing
4. radiographic appearance of the ulnar nonunion is typical with pencilling of the fragment ends—this not uncommon problem results from overprotection by the radius, providing no stimulus for bony healing of the ulna
5. note the return of normal bone density on the last study—the dog was a watchdog and had been returned to work at this time
6. soft tissue swelling seen on the last study may indicate the need to consider removal of the plate and screws
7. observe that screws of similar length may appear to have different lengths on the radiograph because they are placed at different angles to avoid drilling all of the screw holes in the same longitudinal plane, which results in bone weakening

Summary:

In an effort to save a few dollars, delaying and attempting to treat this type of fracture using outmoded methods frequently leads to dissatisfaction. The benefits are negated by protracted healing time and an increase in care required by the owner. It is best to recognize the requirements for fracture treatment clearly, establish the best treatment, and act accordingly. The fracture would heal eventually with an external splint; however, the radiographic appearance of the healing would be markedly different because of the development for a massive, exuberant callus.

Because of the short distal radial fragment with an oblique fracture line, the eventual choice of a plate of proper length and size is excellent. Placement of four screws above and three below was combined with the use of an interfragmentary compressive screw. The plate was placed proximally so as not to interfere with the radiocarpal joint. Synostosis seems to have occurred at the fracture site, and was not the result of the plating. This seems to occur more frequently than is reported.

The ulnar nonunion is clinically unimportant, even in a dog belonging to a large breed. It is presumed to be the result of a large unicorporated fragment. Giant breeds, however, frequently require plating of both bones to provide more adequate support. Note that the quality of bone improved after the use of a bone plate that permitted the dog to bear weight.

CASE 15

Signalment: 7 years old, male, intact, Pit Bull (Staffordshire Terrier)

History: in dog fight with injury to the left forelimb

Physical Examination: marked crepitus and instability of the left radius and ulna

Radiographic Examination: left antebrachium

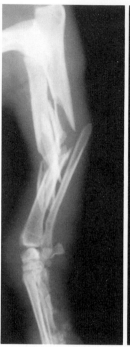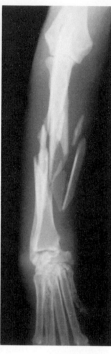

Study at time of entry (day 1):

1. highly comminuted fractures of mid-shaft radius and ulna
2. multiple large butterfly fragments
3. marked limb shortening
4. soft tissue swelling
5. long oblique fracture of fourth metacarpal bone
6. elbow and antebrachiocarpal joint appear normal

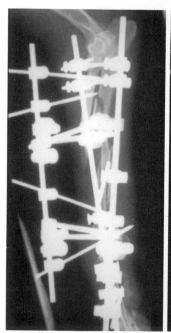

Postoperative study (day 10):

1. reduction and stabilization using a type III K-E apparatus with transfixation pins placed through the radius and metacarpal bones
2. no end-to-end apposition of the fracture fragments
3. caudal angulation of the distal fragments
4. most proximal pin encroaches on the humeroradial joint

Comments on radiographic findings:

1. severity of injury suggests that the condition of the soft tissues should be carefully examined
2. fragment apposition and alignment is difficult to evaluate because of the complexity of the opaque shadows made by this pattern of K-E apparatus

Summary:
This severely fractured radius and ulna was treated with a type III K-E apparatus with three centrally threaded fixation pins, five 1/2 pins with threaded tips, and four 1/2 pins with threaded tips that come off the cranial connecting bar. No reduction of fragments and no apposition of fragment ends was obtained.

It might have been possible to have lagged the fragments and plated the fracture; however, soft tissue injury was severe due to bite wounds. The fact that the dog had been unable to extend the carpus and the fact that the limb was swollen and cold led the surgeon to use a K-E apparatus. Shortly after placement of the K-E apparatus, areas of skin from the left paw began to slough and the left forelimb was amputated.

Treatment of this patient was complicated by the presence of similar fractures on the opposite forelimb. The fractures on that limb were plated and were noted to be healed after 4 months.

Ulna

Signalment: 3 years old, male, intact, Samoyed

History: struck by a car

Physical Examination: limping on left forelimb

Radiographic Examination: left elbow

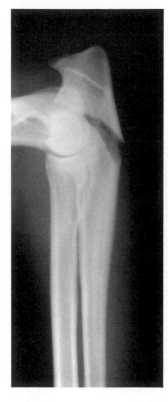

Study at time of entry (single lateral view) (day 1):

1. simple oblique fracture separating the olecranon process with proximal displacement of the fragment
2. fracture line enters the elbow joint

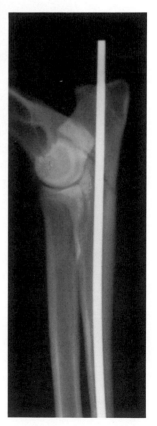

Postreduction study (single lateral view) (day 1):

1. single IM pin extends to the distal third of the ulna, crossing the fracture line
2. good reduction of the fracture fragment

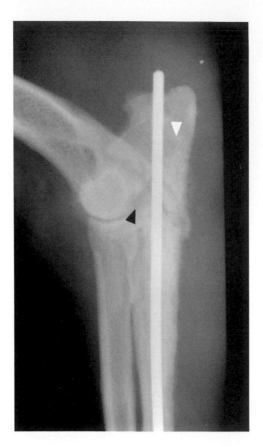

Postreduction study (single lateral view) (day 30):

1. IM pin remains in position
2. fluffy periosteal reaction along length of ulna (osteomyelitis)
3. widening of the fracture line (osteomyelitis)
4. lucencies within the fracture fragment (osteomyelitis) (white arrowhead)
5. joint surfaces not clearly seen radiographically (suppurative arthritis; black arrowhead)

Comments on radiographic findings:

1. this type of fluffy periosteal response in a skeletally immature patient with a fracture may be due to the nature of the periosteal stripping at the time of the trauma—in a patient with severe soft tissue injury, an early healing callus also may appear in this manner; in a patient with an unstable reduction, it may be due to motion at the fracture site and an inefficient deposition of bony callus; in this patient, with soft tissue swelling and inability to use the limb, it is probably due to an infectious process (osteomyelitis)
2. scattered lucencies within the fragment and the ulnar cortex are distant from the fracture site and strongly suggest an infectious process and not callus formation
3. the radiograph made at 4 months posttrauma showed progressive fracture healing, but with more conclusive evidence of osteomyelitis
4. the radiograph made at 8 months posttrauma continued to show fracture healing, clearance of the bone infection, but with severe postinfectious secondary joint disease

Summary:
Accurate reduction is most important in this patient because the fracture enters the joint, causing disruption of the articular surfaces. Joint congruency must be restored for a satisfactory result. The placement of a single IM pin throughout the proximal ulna is not sufficient in itself to maintain stability. In line with current practice, it is understood that the proximal fragment can rotate and may separate, and it is necessary to counter the pull of the triceps muscle. A smaller pin should have been placed slightly cranial to the insertion of the larger pin, avoiding the joint surface. It would have been difficult in this patient, but could have been attempted. In addition, a figure-8 tension band should have been placed caudally.

In this patient, micromotion has led to fragment movement at the fracture site. In addition, bone infection is present, causing a pattern of periosteal response and cortical lucency. The osteomyelitis appears to be soft tissue in origin and is probably iatrogenic. Because micromotion affects the integrity of newly forming blood vessels, it permits an infectious process to progress more readily.

This patient had a relatively easy fracture to repair, but equalizing the pull of the triceps should have been considered to avoid the delay in healing due to fragment motion. Better stabilization might also have influenced the pattern of the infection; certainly, it would have resulted in less severe joint disease.

CASE 2

Signalment: 10 years old, male, intact, Miniature Poodle

History: dog became lame while playing with other dogs in the household

Physical Examination: limping on left forelimb

Radiographic Examination: left elbow

Study at time of entry (day 1):

1. avulsion fracture of the olecranon process with proximal and medial displacement of the fragment
2. fracture appears to be nonarticular
3. elbow joint appears to be radiographically normal

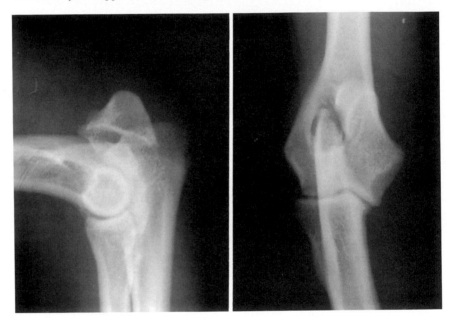

Postreduction study (day 3):

1. tension-band device used to reduce the fracture fragment adequately

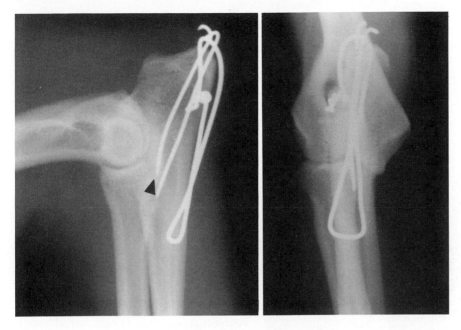

Comments on radiographic findings:

1. determining whether the fracture is articular is important in the prognosis of the injury and the selection of method of repair

Summary:

This is an uncommon type of fracture; most olecranon fractures involve the elbow joint. It is important to consider treatment that results in a tightly held fragment. In this patient, fixation was achieved using two K wires and a figure-8 tension-band device. Note how the shorter K wire tried to enter the more dense bone of the trochlear notch and was reflected caudally (arrowhead). It is not usually necessary to remove the metallic hardware in this patient.

This is a simple, easy, inexpensive way to treat this fracture with an excellent result.

Case 3 **Signalment:** 2 years old, male, intact, domestic long-haired cat

History: struck by a car 24 hours ago

Physical Examination: limping on right forelimb—in shock

Radiographic Examination: right antebrachium (lateral views only)

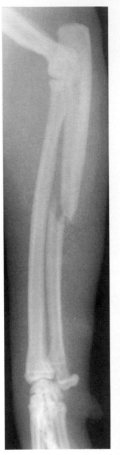

Study at time of entry (single lateral view) (day 1):

1. short oblique fracture of midshaft ulna with displacement of fragments
2. external rotation of foot
3. soft tissue swelling
4. elbow and antebrachiocarpal joints radiographically normal
5. growth plates show a dense bony pattern normally seen shortly after closure

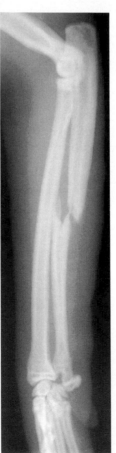

Postreduction study (single lateral view) (day 21):

1. radiographic appearance of the fracture remains as before

Postreduction study (single lateral view) (day 45):

1. radiographic appearance of the fracture remains as before, with only slight modeling at the fracture site

Postreduction study (day 90):

1. sclerosis of the fragment ends with rounding and failure of bridging callus

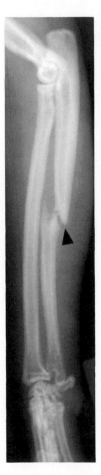

Comments on radiographic findings:

1. typical formation of a nonunion fracture with modeling at the fracture site leading to formation of "elephant foot" pattern (arrowhead)
2. widening of the fracture ends as a pseudoarthrosis forms
3. absence of osteopenia or soft tissue atrophy suggests that the cat continued to use the limb throughout this period of healing, and a rather solid fibrous union probably had united the ulnar fragments

Summary:

The ulnar fracture with intact radius is unusual and probably resulted because the bones in the cat are not as rigid as in the dog. Choices in the treatment of the ulnar fracture could have included: (1) no treatment other than a supportive bandage, (2) placement of a small-diameter IM pin into the ulna from the olecranon to maintain fragment alignment, or (3) use of a plate. Although the election to use no treatment of the fracture risks nonunion or at least delayed union, the use of a plate might be considered an overtreatment.

The cat pronates and supernates its paw frequently, and this motion plus normal activities apparently led to motion at the fracture site and prevented the fracture ends from remaining in apposition. Because the adjacent radius provided the cat with sufficient support for the limb, the requirement for bony healing was not present and the fracture probably "healed" with a fibrocartilaginous union.

Economic considerations unfortunately often determine the choice of reparative procedure to be used. The immediate prospect of limited expenditure in this case led the surgeon to choose the less costly route, but this was a poor choice in the long run.

Carpus–Metacarpus

Signalment: mature, male, intact, Doberman Pinscher

History: dog injured foot while running

Physical Examination: open fracture of left front foot

Radiographic Examination: left front foot

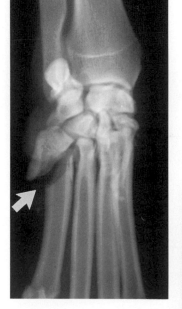

Study at time of entry (single oblique view) (day 1):

1. oblique fracture in the proximal third of the fifth metacarpal bone (arrow)
2. separation of fracture fragments with lateral angulation of the proximal fragment
3. soft tissue swelling supporting clinical finding of an open fracture

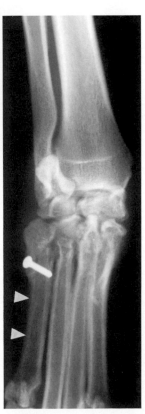

Postoperative study (single oblique view; marked swelling of the foot with draining tract laterally) (day 32):

1. reduction of fracture through placement of single lag screw
2. status of fracture healing difficult to ascertain
3. apposition and alignment of fragments are good
4. note that osteopenia is limited to the fifth metacarpal bone (arrowheads)

177

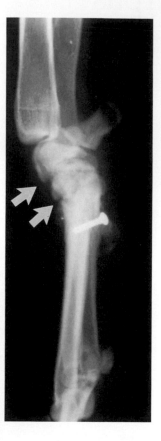

Postoperative study (single lateral view; continued marked swelling of the foot with draining tract laterally) (day 53):

1. single lag screw remains in position
2. destruction of midcarpal and carpo-metacarpal joints (infectious arthritis) (arrows)
3. destructive pattern within carpal bones and proximal metacarpal bones (osteomyelitis)

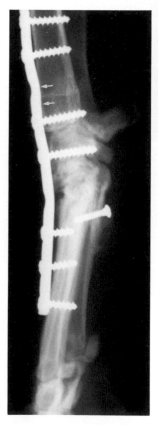

Postoperative study (single lateral view) (day 61):

1. placement of bone plate to achieve arthrodesis
2. collapse of joint spaces

Comments on radiographic findings:

1. early fracture healing with minimal callus formation and disappearance of the fracture line may be difficult to distinguish from bone infection; the clinical history is of greatest importance in making this distinction
2. osteopenia in one bone is not caused by disuse, which would affect all bones equally—a single bone affected is more suggestive of bone infection
3. status of bone infection after the second surgery is difficult to ascertain; surgical debridement of articular cartilage and subchondral bone plus use of large cancellous bone graft makes radiographic evaluation difficult

Summary:
The proximal metacarpal bone fracture was treated with a small lag screw placed to cross the fracture line. The injury was reported as contaminated with a likely chance of infection. With the onset of infection as seen on the first postoperative study, additional treatment should have been instituted. The infection of the carpometacarpal joint resulted in its destruction, and ultimately the treatment was an arthrodesis. The plate was contoured; however, there was a separation between the plate and the proximal metacarpal bones, because of the palmar displacement of the carpal bones. A groove was cut in the cranial aspect of the distal radius to accept the plate (arrowheads). It is interesting to speculate why a screw was not placed through the third screw hole.

The original treatment of the metacarpal fracture was good. Exposure of the fifth metacarpal is relatively easy. However, infection resulted in delayed healing of the fracture and, more important clinically, progressed to an infectious arthritis. The selection of an arthrodesis was good even in the face of the infection. There was little else to do to maintain some stabilization during the healing of the infectious process. A cancellous graft was used to assist in the healing.

Signalment: 4 years old, male, intact, Norwegian Elkhound

History: dog jumped from truck window while truck was moving

Physical Examination: marked hyperextension of right foot

Radiographic Examination: right carpus

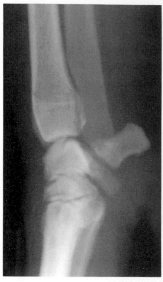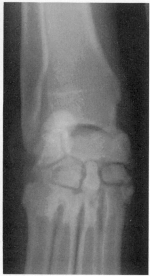

Study at time of entry (routine views) (day 1):

1. no evidence of bone or joint injury

Study at time of entry (stress views both feet) (day 1):

1. marked hyperextension with abnormal motion at carpometacarpal joints of right foot (arrow)
2. normal motion at carpometacarpal joint of left foot

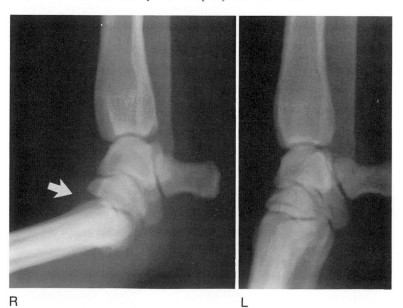

R   L

Postoperative study (day 2):

1. placement of large bone plate and screws across all three carpal joints to achieve arthrodesis

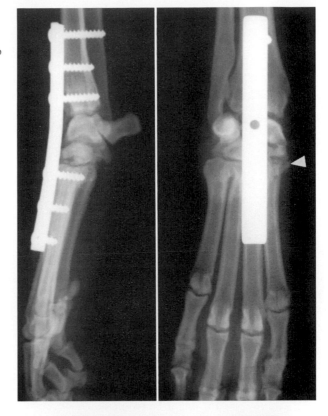

Postoperative study (3 years later):

1. persistent bony ankylosis of radiocar-pometacarpal joints after arthrodesis
2. removal of bone plate and screws occurred 2 years earlier

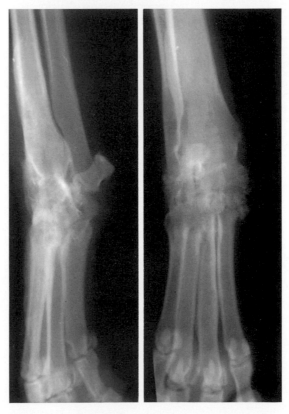

Comments on radiographic findings:

1. note that on the stress study the accessory carpal bone remains in position, indicating injury to the palmar ligaments
2. debridement of articular cartilage and subchondral bone and presence of cancellous graft often is not identified radiographically on the postoperative study—a part of the graft is identified at the carpometacarpal joint medially (arrowhead)
3. pattern of bony ankylosis is as expected
4. osteopenia is more prominent than expected, suggesting a degree of disuse—note thin radial cortices

Summary:

The choice of a panarthrodesis is acceptable; however, with a carpometacarpal subluxation it is possible to spare the antebrachiocarpal joint by fusion of the carpometacarpal joints through placement of IM pins in the third and fourth metacarpal bones into the distal row of carpal bones. The resulting arthrodesis would involve only the carpometacarpal and intercarpal joints.

The surgeon elected to perform a complete arthrodesis using a bone plate. The plate is slightly contoured and fits well against the bone surfaces. Note the ovoid shape of the distal radius with the resulting very small craniocaudal bone diameter, and how the first screw penetrates into the ulna. It is customary to place the plate in such a position that a screw can be placed into the radial carpal bone; however, that was not done in this patient. The angle of the plate shows a slight degree of extension. This angle needs to be based on the breed characteristic. It is helpful to radiograph the opposite limb to determine this. In this Elkhound, the almost upright position is probably correct, whereas, for example, in the German Shepherd Dog, an angle of 10° to 15° might be required.

A lick granuloma that developed 1 year after healing suggests the presence of a loose screw with plate loosening, which would require removal. Owners should be advised that metallic devices may need to be removed even up to months or years after placement and healing. The lick granuloma healed after removal of the plate and screws.

Generalized osteopenia was noted 3 years after the initial injury. This is made especially evident by the thin radial cortices, and is probably due to the dog continuing to favor the limb.

Signalment: 7 years old, male, intact, Doberman Pinscher

History: dog kicked by horse

Physical Examination: open fracture of left foot

Radiographic Examination: left front foot

Study at time of entry (lateral and oblique views) (day 1):

1. severe fracture–luxation of carpometacarpal joints
2. caudal and medial luxation of heads of the second, third, and fourth metacarpal bones
3. fracture of proximal head of second metacarpal bone (arrowhead)
4. open transverse fracture of midshaft of fifth metacarpal bone with fragment separation
5. extensive soft tissue swelling
6. soft tissue gas

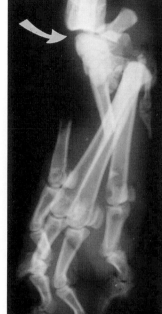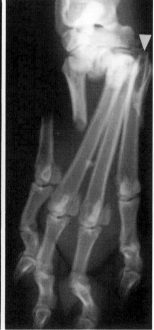

Postoperative study (single dorsopalmar view) (day 2):

1. reduction of luxated third and fourth metacarpal bones with IM pins
2. reduction of fracture–luxation of second metacarpal bone with IM pin
3. plating of midshaft fracture of fifth metacarpal bone
4. apposition and alignment of fragments are good

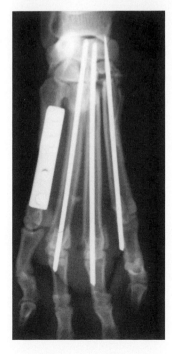

Comments on radiographic findings:

1. note that the apparent laxity of the antebrachiocarpal joint on the lateral view (curved arrow) was due to the injury in the foot and not injury at that site
2. difficult to determine the source of the small triangular opaque fragment within the soft tissues, especially in the foot—this may be a bony fragment, but it is far from the fracture site, and the sharp edges suggest a foreign body such as glass or metal
3. status of the distal row of carpal bones is difficult to ascertain

Summary:

This is an interesting fracture because it was open and showed such severe deformity. Exposure to plate the fifth metacarpal bone was made easier because of the extensive superficial soft tissue injury. If the tissues had been closed, the surgical exposure would have been more complicated. Intramedullary pins were introduced from the distal ends of the metacarpal bones through the joint spaces, penetrated the adjacent carpal bones, and were seated in the radial carpal bone. The pins should be removed immediately after fracture healing to avoid injury to the articular surfaces of the metacarpophalangeal joints.

The possibility of limited motion of the affected joints cannot be avoided. Movement at the antebrachiocarpal joint has been preserved. This joint provides the greatest degree of motion.

A superior technique for insertion of IM pins is to drill angled holes in the dorsal surfaces of the distal metacarpal metaphyses so that flexible pins can be inserted in a proximal direction. The ends are bent upward and cut short, which will protect the metacarpophalangeal joint and provide a nice "handle" that permits rotation and retraction after healing.

Signalment: 4 years old, male, intact, Golden Retriever

History: dog appeared acutely lame while working in field trial competition

Physical Examination: refuses to bear weight on left front foot

Radiographic Examination: left front foot

Study at time of entry (day 1):

1. oblique articular fracture in the proximal end of the third metacarpal bone
2. minimal separation of fracture fragments
3. soft tissue swelling

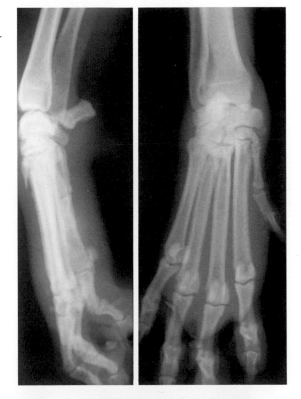

Postoperative study (day 2):

1. placement of a single wire loop to reduce the fracture and include a portion of the palmar ligament

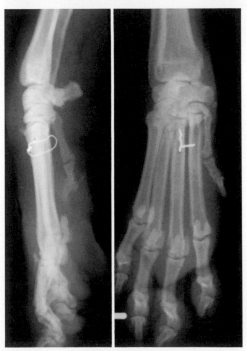

Comments on radiographic findings:

1. soft tissue swelling is generalized and provides little assistance in locating the bone injury
2. injury of this type is difficult to evaluate radiographically because of the overlapping of the heads of the metacarpal bones; multiple views, some made with a stress technique, are used to assist with this problem—stress views made in this dog demonstrated only a minimal degree of hyperextension of the foot

Summary:

This fracture is similar to the carpal chip fractures in racing horces and is probably caused by hyperextension of the foot and a "pinching" action that breaks the corner of the metacarpal bone. The fracture is articular and the pain is severe with movement. The fracture fragment either needs to be removed or stabilized. The choice of a single cerclage wire is a good one, with the wire reducing the fragment and including a portion of the thick palmar ligament. Note how nicely the ends of the cerclage wire are twisted, cut, and flattened against the bone surface so as not to cause irritation within the soft tissues. It is necessary to provide some external support for the foot during early healing.

Removal of the fragment might have been considered, but it would be more difficult. Any resulting void in the metacarpal bone would ultimately have healed.

Signalment: 3 years old, male, intact, Springer Spaniel

History: dog sustained unknown trauma

Physical Examination: refuses to bear weight on left forelimb—foot swollen and painful

Radiographic Examination: left forefoot

Study at time of entry (single dorsopalmar view) (day 1):

1. oblique fractures at junction of proximal and middle thirds of third, fourth, and fifth metacarpal bones (arrows)

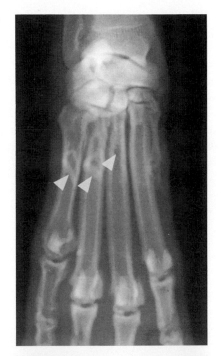

Postoperative study (single dorsopalmar view) (day 2):

1. fracture reduction through use of small bone plates in fourth and fifth metacarpal bones
2. use of a single lag screw in third metacarpal bone
3. apparent good reduction

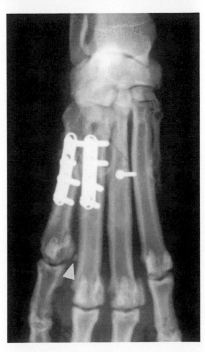

Comments on radiographic findings:

1. the fractures appear radiopaque in the fourth and fifth metacarpal bones (arrowheads) because of superimposition of the fracture fragments, whereas the fragments are separated in the third metacarpal bone, causing the fracture line to appear radiolucent (arrowhead)
2. chronic injury at the fifth metacarpophalangeal joint seems to involve the seventh sesamoid bone as well (arrowhead), but is unrelated to the acute trauma
3. remember that multiple views are necessary to ascertain the exact character of fractures
4. the oval nature of the screw hole plus the small size of the round screw head causes a lucent ring and erroneously suggests that the screw is not tightened

Summary:
The use of finger plates (toy plates) is an excellent choice, as is the use of a small screw to lag the third metacarpal fracture. The level of support for healing is adequate with the usual additional external splinting. Healing of fractures treated in this manner usually is rapid, and the splint can be removed within 3 to 4 weeks. Usually, the plates need not be removed unless complications arise, such as loosening of the screws.

Signalment: 5 months old, male, intact, Siberian Husky

History: puppy caught right front foot in spokes of owner's bicycle

Physical Examination: refuses to bear weight on right forefoot

Radiographic Examination: right forefoot

Study at time of entry (day 1):

1. transverse fractures (with "stairstep") of midshaft of second, third, fourth, and fifth metacarpal bones
2. marked lateral and dorsal angulation of distal fragments
3. little end-to-end apposition of fragments
4. soft tissue swelling
5. physeal plates normal for puppy of this age

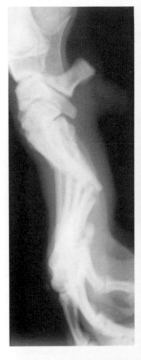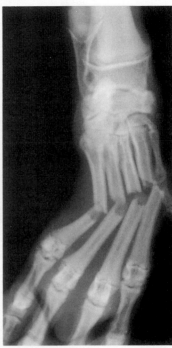

Postoperative study (day 2):

1. excellent fracture reduction through use of small bone plates on the third and fourth metacarpal bones
2. second and fifth metacarpal bone fractures left untreated

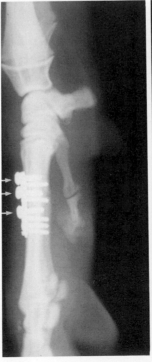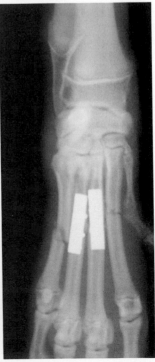

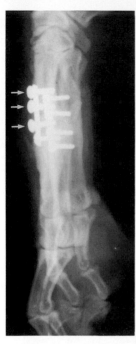
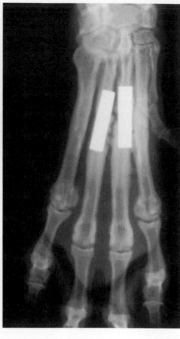

Postoperative study (day 60):

1. plate and screw position as before
2. fracture reduction as before, with healing in all fractures
3. note closure of physeal plates, as expected

Comments on radiographic findings:

1. note the closure of the distal metacarpal physeal plates; evaluation of the opposite limb would be necessary to determine whether this was hastened by the trauma
2. note that in the plate in the third metacarpal bone the three proximal screws are larger, the heads project, and they appear differently

Summary:

One of the superior options available to the surgeon is placement of finger plates, as in this patient. Placement of the plates and screws was excellent in treatment of the low energy fracture. Note that five screws are smaller and seat nicely in the plates, whereas three screws are slightly larger and their fit in the plates was less perfect (arrows). Note how use of plates to reduce the fractures in the third and fourth metacarpal bones brought the fragments in the other two bones into nice alignment. Healing occurred with excellent apposition and alignment of the fragments. The foot required support through external mobilization during early healing because of the two unsupported fractures. The expected healing time for these fractures would be 4 weeks, considering the age of the patient.

It also would have been acceptable to introduce IM pins distally to reduce and fix all four of the fractures; however, use of bone plating is superior.

Femur

Signalment: 4 months old, female, intact, Labrador Retriever

History: dog seen to be hit by a car

Physical Examination: dog presented in shock with abrasions on abdomen and probable fracture of right femur—abdominal paracentesis was negative

Radiographic Examination: right femur

Study at time of entry (day 1):

1. oblique comminuted fracture of femur at junction of proximal and middle thirds
2. several displaced bufferfly fragments
3. single linear fracture line extending distally into distal fragment (arrow)
4. marked displacement with overriding of fracture fragments
5. radiographically normal hip and stifle joint
6. growth plates as expected for a 4- to 6-month old puppy

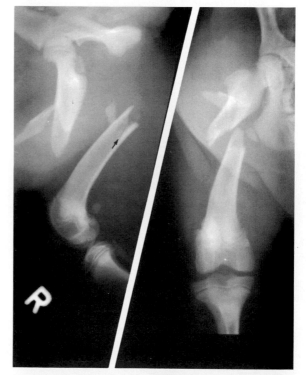

Postoperative study (day 17):

1. reduction using a single IM pin with three cerclage wires
2. apposition and alignment of fragments near-anatomic
3. marked retroversion of femoral head and neck
4. callus formation typical for puppy, with "bucket handle" appearance (arrows)
5. fracture line partially visualized

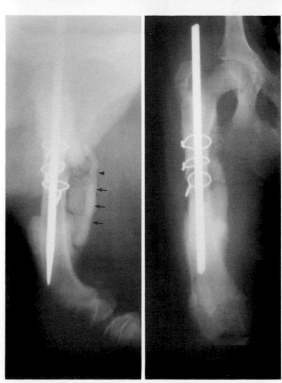

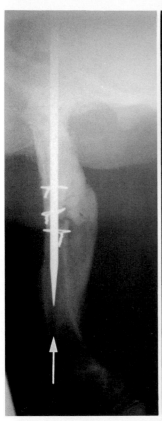 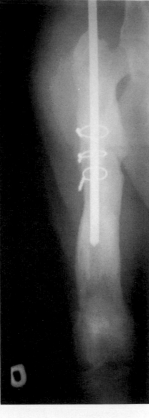

Postoperative study (day 32):

1. IM pin is "backing-out" with lessening contact within the distal fragment (note the pin track; arrow)
2. cerclage wires remain in position
3. callus formation primarily caudal and medial
4. cranial angulation of distal fragment
5. persistent retroversion of femoral head and neck
6. fracture line partially visualized
7. bone length increased 1.5 cm

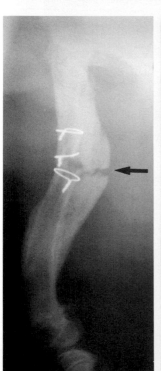 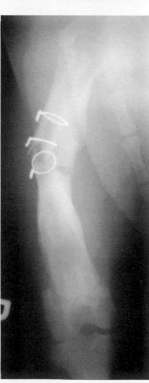

Postoperative study (day 36):

1. refracture of femur after IM pin removal
2. fracture of caudomedial callus (arrow)
3. cranial and medial angulation of the distal fragment
4. femoral head and neck retroversion as before

Second postoperative study (day 60):

1. use of contoured plate and screws to repair refracture (note the nut and washer on the distal screw)
2. good apposition and alignment of fragments
3. further femoral head and neck retroversion
4. retention of small piece of cerclage wire
5. 2.3-cm increase in femoral length due to bone growth
6. note the proximal position of the patella

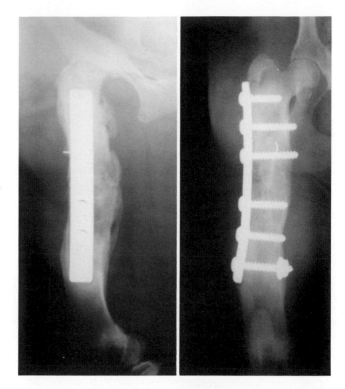

Postoperative study (day 140):

1. fracture healing
2. excellent modeling with restoration of medullary cavity
3. retroversion of the femoral head and neck not as completely evaluated
4. plate and screws remain in position
5. piece of wire as before
6. note the proximal position of the patella

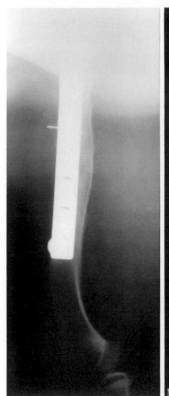 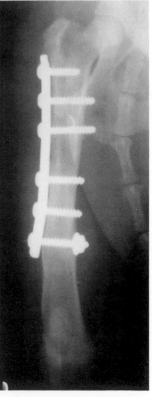

Comments on radiographic findings:

1. pelvic fractures are not clearly seen on these studies, but were more easily evaluated on a separate pelvic radiographic study
2. on the 1-month study, note the avascular nature of the craniolateral fragment at the fracture site—this plus visibility of the fracture line suggests delayed healing
3. the presence of delayed healing did not prevent the decision to remove the loosened IM pin, with resulting refracture
4. note the marked increase in length of the femur after the second surgical repair
5. the position of the femoral head and neck are not easily evaluated on femoral views—an additional lateral view of the hip joint would provide this information
6. the proximal position of the patella interferes with formation of the femoropatellar joint by preventing formation of the deep trochlear notch in the distal femur

Summary:

This is a nasty fracture with comminution and linear fractures altering the character of the injury. The caudal position of the distal fragment with marked overriding is typical for midshaft femoral fractures. Open reduction was used to place a chisel-pointed IM pin. The pin is small, does not fill the medullary cavity, and the distal point should have been inserted more deeply. The surgeon should avoid following the cranial cortex with the pin. Following the caudal cortex will result in a deeper and more solid placement of the pin. The distal cerclage wire held, but was located very close to the fracture site. The levering effect of the distal fragment on the poorly located pin moved it proximally, allowing the distal cerclage wire to fall into the fracture site. The loosened pin had to be removed because it was protruding, not supporting the fracture, and penetrating the skin at the dorsal point of insertion. The callus bridged caudally and medially, but was not strong enough to support the bone after removal of the pin.

The subsequent fracture of the partially healed bone was treated with a well contoured heavy plate and three screws proximally and three screws distally. Note the use of cancellous screws in this immature bone. Even then, a nut and washer was needed on the distal screw because it stripped its threads. A longer plate could have been used. Unfortunately, the retroversion of the femoral head could not be corrected at the time of the plating. Anatomic structures are difficult to recognize when the fracture site is surrounded by fibrocartilaginous callus.

In this dog, the consequences of improper IM pin insertion were corrected by the adequate placement of a bone plate, which was followed by good healing. Neither the retroversion nor the possible damage to the femoropatellar joint were addressed, which may have been due to failure of observation. The proximal position of the patella suggests possible development of quadriceps contracture. Fractures at the age of 4 months are particularly troublesome because of the exuberant callus formation due to periosteal stripping and the intense growth activity.

Signalment: male, intact, domestic short-haired kitten

History: the kitten was found by neighbors and brought to the clinic with a fractured left femur

Physical Examination: marked swelling of left pelvic limb, subcutaneous emphysema over thoracic region, inguinal laceration

Radiographic Examination: left femur

Study at time of entry (day 1):

1. midshaft oblique fracture of femur with single small butterfly fragment
2. overriding of fragments
3. hip and stifle joints appear radiographically normal
4. physeal plate appearance typical for 4- to 6-month old kitten

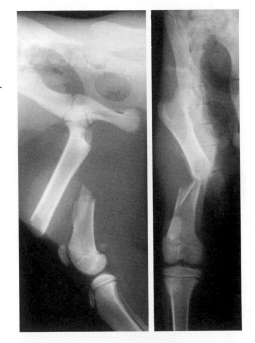

Postoperative study (day 3):

1. placement of two smooth, stacked IM pins
2. use of two cerclage wires distally to hold the butterfly fragment in position
3. apposition and alignment of fragments appear good
4. subcutaneous emphysema

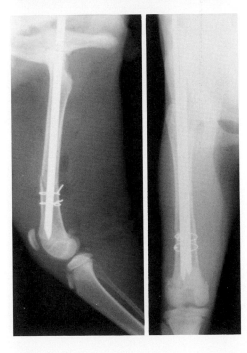

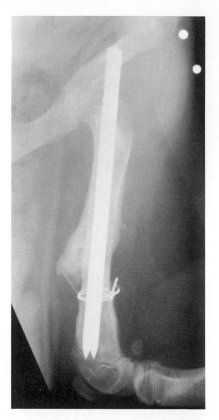

Postoperative study (single lateral view) (day 28):

1. heavy callus formation on proximal fragment
2. apparent withdrawal of IM pins
3. slippage of cerclage wires
4. cortical fragments identified at fracture site
5. no callus bridges the fracture site

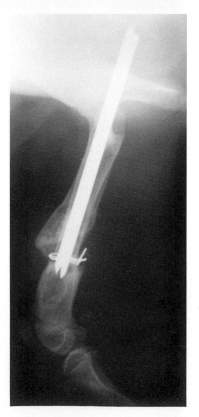

Postoperative study (single lateral view) (day 42):

1. heavy callus formation on proximal fragment
2. marked withdrawal of IM pins
3. cerclage wires remain repositioned at the fracture site
4. distal cortical fragment shows no callus formation
5. no callus bridges the fracture site

Postoperative study (day 190):

1. malunion fracture
2. removal of IM pin
3. cerclage wires buried in callus
4. marked lateral angulation of distal fragment
5. retroversion of femoral head
6. near closure of physeal plates

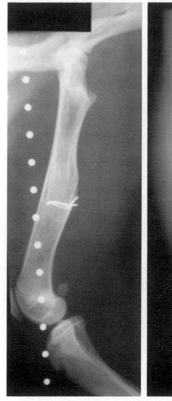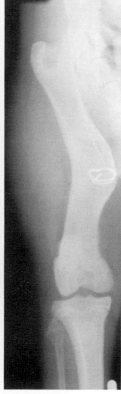

Comments on radiographic findings:

1. there is some question of whether the appearance of the bone should have been evaluated as an osteomyelitis—it probably was only excessive motion at the fracture site associated with slippage of the cerclage wires and creation of a fulcrum point—note how the movement of the cerclage wires destroys the entering capillary bed, leaving the distal fragment avascular
2. heavy opaque "dots" on two radiographs mark 1-cm distances and are of value in determination of bone length when using a magnification technique
3. note the neocortex formed laterally and cranially as a result of the angulation—the wires can be used to determine the original location of the cortical fragments
4. last lateral radiograph is best positioned, and the retroversion of the femoral neck can be accurately determined

Summary:
The use of stacked IM pins in treatment of a midshaft femoral fracture seemed logical to the surgeon. However, in this midshaft fracture the pull of the large flexor muscles caused leverage of the distal fragment, which drove the pins proximally. The cerclage wires moved into the fracture site due to the motion of the fragments. The surgeon had to pull the IM pins, and chose to observe the progress of the fracture. Although the bone healed, unfortunately a malunion resulted. The level of angulation of the distal fragment, although important in the dog, will be compensated for rather well by the cat. At the time of removal of the IM pins, it would have been possible to use an external fixator with three pins proximally and two pins distally (3/2), or apply a small bone plate—either of which would have prevented the malunion.

CASE 3 | **Signalment:** 3 months old, male, intact, domestic short-haired kitten

History: kitten found crying

Physical Examination: left pelvic limb crepitus

Radiographic Examination: left femur

Study at time of entry (day 1):

1. midshaft spiral fracture femur with butterfly fragments
2. fracture lines extend proximally and distally
3. minimal overriding of the fragments
4. hip and stifle joints appear radiographically normal
5. physeal plates appear typical for kitten of this age

 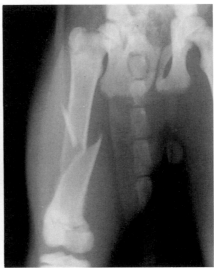

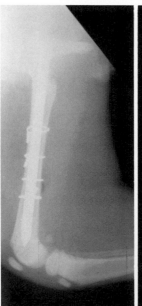 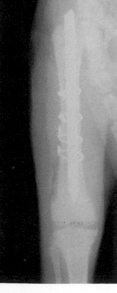

Postoperative study (day 2):

1. placement of a single, large, smooth IM pin
2. use of four cerclage wires
3. apposition and alignment of fragments appears good
4. subcutaneous emphysema

Postoperative study (day 24):

1. heavy callus formation at fracture site proximally
2. callus does not bridge the distal oblique fracture lines
3. fracture lines still identified
4. cerclage wires buried and in same position as before
5. lengthening of femur due to bone growth

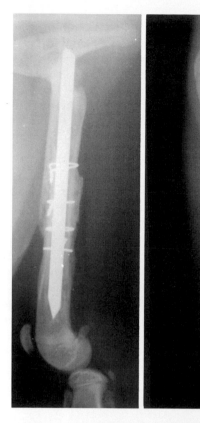

Postoperative study (day 38):

1. callus maturing
2. fracture lines not seen
3. cerclage wires buried and in same position as before
4. apposition and alignment appear as before
5. continuing bone growth

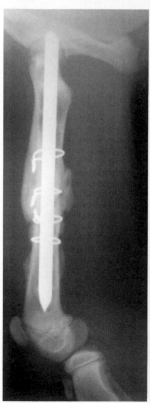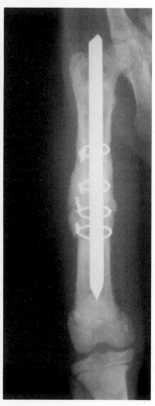

Comments on radiographic findings:

1. spiral character of the fracture carries the fracture lines proximally and distally into both metaphyseal regions of the femur (arrows); determining the extent of fracture lines obviously is important
2. appearance of the fracture changes quickly in an immature patient—the last study shows healing, whereas the study 14 days earlier showed only questionable callus formation

Summary:

The feline femur is straight and accepts a relatively large-diameter IM pin. It is easy to find the trochanteric depression in the proximal fragment and direct it in a normo-grade direction. The four cerclage wires were used to retain the fracture fragments. The possibility of rotation was not a problem because of the obliquity of the fracture line.

The cat usually will remain quiet during fracture healing, permitting the use of less rigid fixation. The femur lengthened during healing due to growth, which often is accelerated because of the hyperemia induced by the trauma.

The proximal end of the pin may be reached and its removal made easier with a "root-searching" dental forceps. The muscle mass over the pin tip in the cat is much less than that in the dog, allowing for an easier approach.

This is a good example of a satisfactory treatment of a midshaft femoral fracture in a cat with the use of a rather large IM pin. The use of four properly applied cerclage wires was very important in the stabilization of the fragments, and resulted in a nicely healed fracture.

Signalment: 2 years old, female, intact, mixed breed dog

History: dog found dragging left pelvic limb

Physical Examination: crepitus in stifle region

Radiographic Examination: left femur

Study at time of entry (day 1):

1. note generalized osteopenic bone with thin cortical shadows
2. fracture at junction of middle and distal thirds of the femur
3. impaction at fracture site
4. comminution with single fracture line extending proximally
5. physeal plates typical for dog 4 to 6 months of age, suggesting reported age is incorrect or the patient has generalized hormonal disease delaying physeal closure
6. hip and stifle joints are radiographically normal

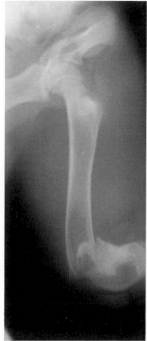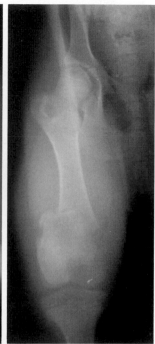

Postoperative study (day 4):

1. reduction of fracture using a single smooth IM pin with nonanatomic reduction
2. pin seated distally within epiphysis cranially
3. pin proximally within the caudal and lateral aspect of the femoral head (arrows)
4. apposition and alignment of fragments are good, with minimal lateral displacement of the distal fragment
5. minimal anteversion of femoral head

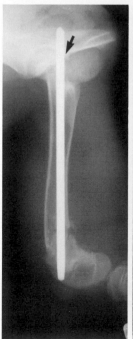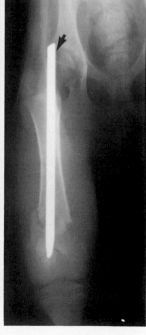

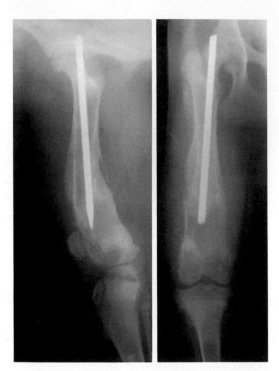

Postoperative study (day 43):

1. healing of the fracture
2. relative movement of the IM pin proximally due to bone growth (0.5 cm)
3. proximal tip of the pin remains near the hip joint
4. apposition and alignment of fragments are good
5. anteversion of the femoral head re-mains
6. stifle joint hyperextension suggests that the limb has been maintained in complete extension during fracture healing

Comments on radiographic findings:

1. it is important to note the osteopenia and classify this as a pathologic fracture because healing will be complicated
2. impaction of fragment ends is uncommon in healthy bone—usually, the frag-ments are separated, with the distal fragment displaced caudally and proximally
3. the importance of noting the location of the IM pin proximally is illustrated in this patient
4. the effect of growth of a long bone can be seen by noting the shift in location of both the proximal and distal tips of the IM pin
5. it is difficult to determine disuse osteopenia superimposed over the generalized osteopenia present at the time of injury

Summary:

It is important to note that this is not the usual radiographic appearance of a distal femoral fracture, but rather of a pathologic fracture in osteopenic bone. The cause of the generalized osteopenia may be dietary or possibly associated with phosphorus retention due to renal disease.

The surgeon chose to use a single IM pin with a chisel point. The pin was inserted in a retrograde manner and exited very close to the femoral head. One of the hazards of retrograde pin introduction is lack of control of its exit point. This hazard can be reduced if the pin is inserted normograde.

The fracture healed, but use of the leg was severely affected by injury to the femoral head. Subsequent alteration of the location of the femoral condyles resulted in a change in the angle of the stifle joint. Because of the hyperextension of the joint, the patella is positioned more proximally than normal. This plus the injury to the femoral head resulted in permanent mechanical lameness.

Signalment: 7 months old, male, intact, Afghan Hound

History: dog found by side of road, unable to walk

Physical Examination: marked increase in right lung sounds; crepitus in right femur

Radiographic Examination: right femur

Study at time of entry (day 1):

1. long oblique fracture at junction of proximal and middle thirds of femur
2. lucent line extending distally in distal fragment, probably nutrient vessel channel
3. marked overriding of fracture fragments
4. radiographically normal hip and stifle joints
5. growth plates open, as expected in an animal of this age

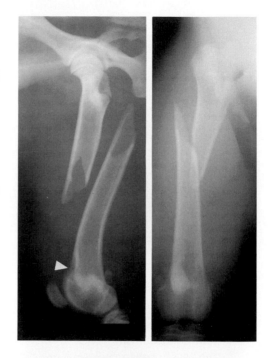

Postoperative study (day 4):

1. reduction using one full cerclage wire, one hemicerclage wire, and three stacked IM pins
2. good apposition and alignment of fragments
3. postsurgical soft tissue gas

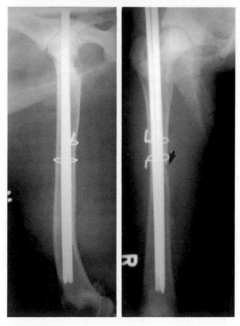

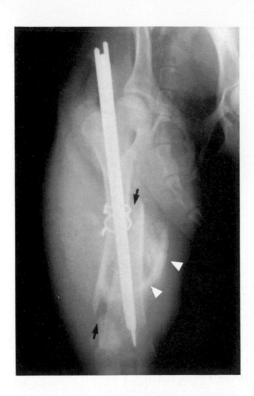

Postoperative study (single craniocaudal view) (day 14):

1. marked retraction of the IM pins
2. cerclage wires remain
3. marked overriding of fragments with bone shortening
4. new fracture line extending proximo-medially to distolaterally within the distal fragment (arrows)
5. apposition and alignment of fragments markedly changed
6. "bucket handle" callus forming caudally and medially (arrowheads)

Comments on radiographic findings:

1. the apparent defect within the cranial femoral cortex just proximal to the patella (arrowhead) is a developmental defect not related to the trauma, and will tend to model with age
2. the cause of the radiolucent line extending into the distal fragment cannot be definitely determined, and thus it should be carefully evaluated at surgery for a fissure fracture line—the fact that a new fracture occurred suggests that the line was a fissure line that spiraled distally
3. note the defect in the cortex within the cortical fragment (arrow); this could have caused weakness at the fracture site
4. the unique pattern of callus formation is secondary to periosteal tearing in a young dog
5. note the suggestion that the cortical bones in this breed are thinner than in other large breeds, and they seem to fracture more readily (a powerful engine in a light body)

Summary:

The surgery was wisely delayed until the dog's condition stabilized. The fracture was nicely reduced using cerclage wires and three stacked pins. Note how the dorsally located hemicerclage wire was used to hold the pins against the caudal cortex. Possibly the fracture was more extensive than was recognized, and a new fracture line formed distally. With the new fracture, impaction occurred and pins were forced proximally. From the appearance of severity of the subsequent fracture, it would seem that it must have been caused by another unreported traumatic episode. With this young, active animal, it is not difficult to presume that restraint was difficult.

Under ordinary circumstances, the use of IM pins and cerclage wires might have sufficed. In this large active dog, it would have been appropriate to have lagged the fracture fragments and placed a bone plate in a neutralization mode. However, even with plating it would have been most necessary to restrain the activities of the dog.

Signalment: 8 months old, female, spayed, domestic short-haired cat

History: attacked by dog, with injury to the left pelvic limb

Physical Examination: crepitus left femur

Radiographic Examination: left femur

Postoperative study at time of entry (day 1):

1. oblique fracture of midshaft femur reduced and stabilized by use of an IM pin and type 1 K-E apparatus with one fixation pin proximally and one distally
2. good apposition and alignment of fragments
3. presence of a Penrose drain (arrows)
4. physeal plates as expected in a patient of this age
5. hip and stifle joints radiographically normal

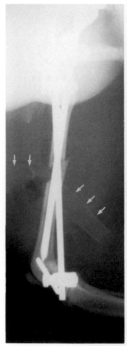
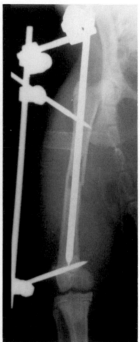

Postoperative study (day 32):

1. minimal stabilization using IM pin and type 1 K-E apparatus, with loosening of both fixation pins
2. heavy callus formation around the fracture line with no bridging noted
3. sequestra identified at the fracture site (arrows)
4. marked cortical lucency suggesting infection in the medullary cavity
5. minimal cranial angulation of the distal fragment

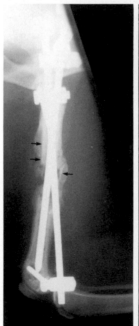
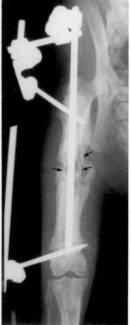

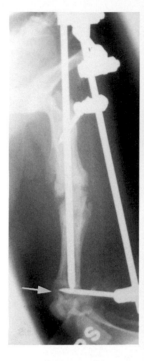 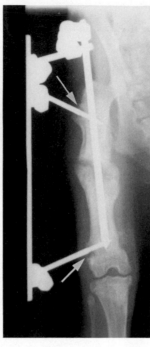

Postoperative study (day 53):

1. poor stabilization using IM pin and type 1 K-E apparatus, with retraction and loosening of the fixation pins
2. sequestra still evident with pockets of lucency at the fracture site
3. heavy callus formation around the ends of the fracture fragments without bridging, suggesting impending non-union
4. minimal cranial angulation of the distal fragment
5. marked soft tissue atrophy

Comments on radiographic findings:

1. note how the IM pin is left long and is attached by a connecting bar to the long connecting bar of the K-E apparatus
2. the shadow of the radiopaque drain must not be confused with a bone fragment or foreign body
3. the sequestra are easily identified at 1 month
4. note the marked lucency around the tips of the fixation pins due to pin motion

Summary:
A midshaft femoral fracture in a kitten was handled initially using an IM pin with a 1/2 K-E apparatus, with the additional feature of attaching a connecting bar between the K-E apparatus and the IM pin. This prevents the IM pin from exiting and adds stability to the device. The Penrose tubing was used as a drain for the soft tissue infection caused by the dog bite. With plating interdicted because of the infection, the surgeon provided the next-best fixation possible. Such cases constitute a race between stabilization and healing of the fragments and the negative effects of the infection. In this patient, the infection won.

In addition, the cat had jumped from a window 10 days after the original fixation. This may have contributed to the loosening of the fixation pins, but obviously did not cause the osteomyelitis. The osteomyelitis appears to be centered at the fracture site, involves the medullary cavity, and has localized in the fixation pin sites. By 1 month, the fixation pins were already loose with marked lucency around the tips, and sequestra were easily identified at the fracture site.

On the last study, nonunion is noted, with rounding of the callus at the fracture site. The pins are retracting and the healing process is in trouble. The problem began with a contaminated wound since there was a history of being attacked by a dog. The stabilization of the fracture was correct, but may have been altered by the jump from the window. Perhaps antibiotic therapy, both local and systemic, might have been used more aggressively. There also may have been a problem of owner cooperation in control of the cat as well as in the administration of the antibiotics.

Because of the osteomyelitis and resulting nonunion fracture, the owner chose euthanasia over amputation.

Signalment: 10 months old, male, intact, domestic long-haired cat

History: cat was outside during the night and found this morning unable to walk

Physical Examination: toenails splintered on right pelvic limb; painful on physical examination; full urinary bladder palpated

Radiographic Examination: right femur

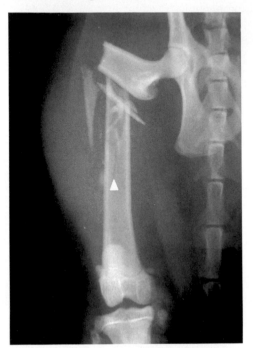

Study at time of entry (craniocaudal view only) (day 1):

1. comminuted fracture at junction of proximal and middle thirds of femur
2. two prominent butterfly fragments
3. linear fracture line extending distally within the distal fragment (arrowhead)
4. marked displacement of fracture fragments
5. radiographically normal hip joint
6. radiographically normal stifle joint with marked soft tissue calcification medial to the joint that is probably a posttraumatic change due to other trauma
7. growth plates open as appropriate for age

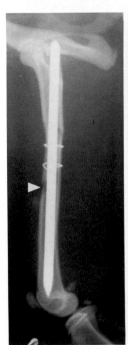
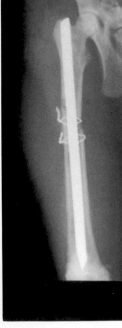

Postoperative study (day 2):

1. reduction using a single smooth IM pin and two cerclage wires with cortical defect cranially (arrowhead)
2. good apposition and alignment of fragments with minimal cranial angulation of the distal fragment
3. large cortical fragment held in position by cerclage wires
4. marked anteversion of femoral head and neck
5. postsurgical soft tissue gas

Postoperative study (day 26):

1. IM pin remains as before
2. callus formation trying to bridge the fracture site
3. fracture line remains visible
4. impaction of fragments with approximately 1.5 cm shortening
5. avascular fragments at fracture site without bony response (arrowheads)
6. apposition and alignment of fragments as before

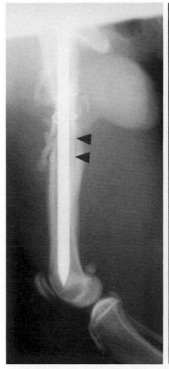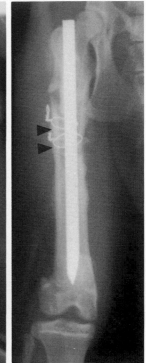

Postoperative study (day 62):

1. IM pin as before
2. mature callus formation bridging fracture site
3. fracture line not visible
4. more severe anteversion of the femoral head
5. growth has compensated for most of the femoral shortening

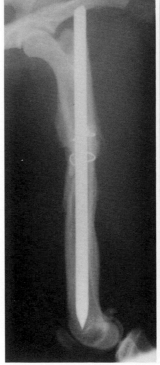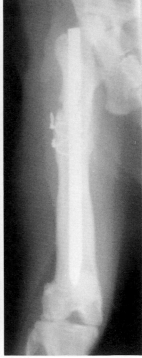

Comments on radiographic findings:

1. if length of bone is important, lateral radiographs need be made of the opposite limb to ascertain normal length
2. note the avascular nature of the cortical fragments at the fracture site and their influence on fracture healing
3. the position of the femoral head and neck can be evaluated only on lateral radiographs in which the femoral condyles are superimposed; note that the second study did not include the hip joint

Summary:

The surgeon chose placement of a large IM pin in treatment of this femoral fracture. The fragments were well positioned through use of the cerclage wires. The wires are relatively small in diameter, more like surgical suture wire than orthopedic wire. Note the defect in the cranial cortex, which may signify a weakness that could have benefited from application of a cancellous bone graft. Unfortunately, the position of the femoral head was ignored and it was left in marked anteversion.

The impaction at the fracture site was unfortunate. This instability resulted in destruction of the developing extraosseous blood supply, loosening any callus that had begun to bridge, and permitted further anteversion of the femoral head. Movement of the cerclage wires was especially destructive of the external blood supply to the developing callus. After the fragments impacted, their motion was controlled and healing proceeded. In addition, the extracortical new bone noted distally was probably secondary to subperiosteal hemorrhage resulting from fragment collapse.

The patient recovered with some degree of lameness due to the femoral head anteversion, and further follow-up was not possible because the cat died from lymphosarcoma 4 months after the injury.

Signalment: 8 months old, male, castrated, Siamese cat

History: hit by car, large swelling and hematoma in left thigh

Physical Examination: left pelvic limb crepitus; soft tissue swelling

Radiographic Examination: left femur

Study at time of entry (day 2):

1. spiral fracture in the midshaft of the femur with butterfly fragments
2. overriding of the fragments
3. hip and stifle joints appear radiographically normal
4. physeal plates appear typical for 8-month-old cat

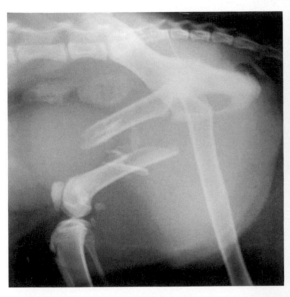 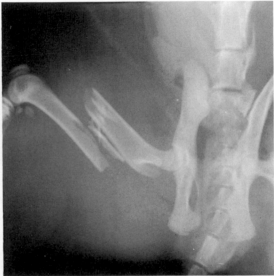

Postoperative study (day 4):

1. placement of a single, large, smooth IM pin
2. cancellous graft laterally (arrows)
3. interlocking of major fragments with shortening of the femoral length
4. apposition and alignment of fragments appears good, with cranial angulation of distal fragment
5. large bone fragment medial to midshaft (arrow)
6. wire skin sutures present

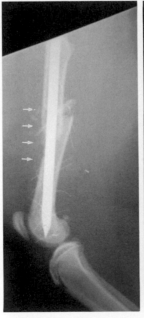 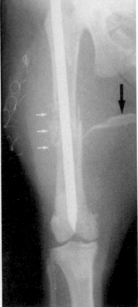

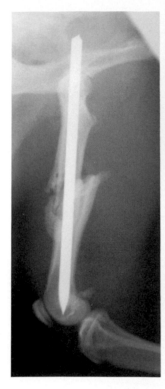
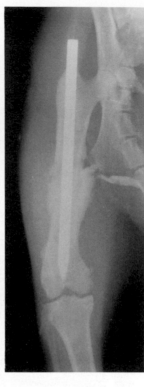

Postoperative study (day 34):

1. single IM pin as before
2. heavy callus formation at fracture site cranially
3. further impaction at fracture site with 1 cm additional shortening of the femur
4. fracture lines still evident
5. bone fragment remains unchanged in soft tissues
6. retroversion of femoral head

Postoperative study (day 68):

1. callus bridges fracture site and is maturing
2. fracture lines not seen
3. apposition and alignment appear as before
4. bone fragment in soft tissue losing density (arrow)
5. bone growth 0.5 cm in length

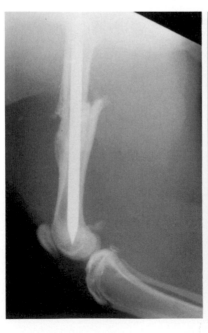
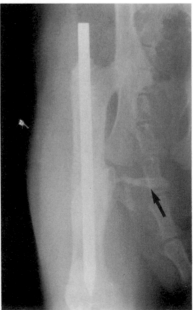

Comments on radiographic findings:

1. note the cortical defect caused by the creation of the butterfly fragment (arrow)
2. retroversion of the femoral head was not seen on the first postoperative study because the hip joint was not included on the radiograph—both proximal and distal joints should be included on all radiographs of fracture cases
3. in this growing patient, the length of the fractured bone may shorten as a result of impaction at the fracture site and may then lengthen due to growth at the physeal plates; in a patient with expectations for high performance it is usual to monitor normal bone length by radiography of the opposite limb

Summary:
The surgeon elected to pin this fracture with a single, deeply seated IM pin. The two major fragments were wedged together. The sacrifice in length of the bone is not important in this patient. It is possible to lose 2 to 3 cm in the length of a long bone in a large dog; for which the dog compensates by altering the angle of the knee. This is not an acceptable sequela, however, in a straight-legged dog (Boxers and Chow Chows).

The smooth pin was inserted in a normograde manner and passage was maintained in the center of the medullary cavity. With a smooth pin, muscular stress can eventually rotate the fracture fragments. A good rotational alignment at the time of reduction may in time fail, due to resorption of the tissues around the pin permitting rotation between the pin and the bone.

Retroversion of the femoral head and neck was not clearly noted until the study at 34 days because of problems in patient positioning. Retroversion is not as important in the cat as in the dog because of limited physical demands.

A dog would have required an external K-E device with the placement of two connecting pins proximally and two connecting pins distally (2/2) to treat a fracture of this type. In the cat, the IM pin fills the medullary cavity, and there is not much room for fixation pins. Because of the cranial location of the distal tip of the pin in the dog, there is adequate space for a transcondylar pin. This situation does not apply to the cat.

Healing occurred in this patient, indicating that a total anatomic reduction is not absolutely necessary, however nice it might look on the postoperative radiograph.

┌─────────┐
│ CASE 9 │ **Signalment:** 1 year old, female, intact, domestic short-haired cat
└─────────┘

History: cat brought into emergency service unable to bear weight on right pelvic limb

Physical Examination: femoral fracture

Radiographic Examination: right femur

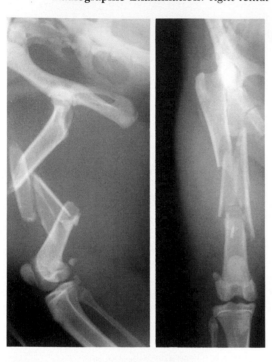

Study at time of entry (day 1):

1. midshaft oblique segmental fracture of femur with longitudinal splitting of segmental fragment into three smaller fragments
2. overriding of fragments
3. hip and stifle joints appear radiographically normal
4. growth plate development commensurate with age

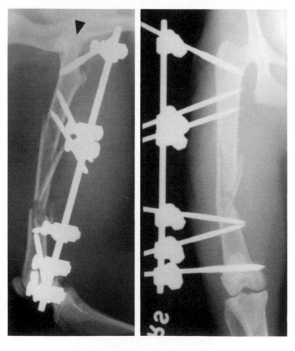

Postoperative study (day 2):

1. placement of a type I external K-E apparatus with three pins proximally and three pins distally (3/3)
2. note the proximal and distal screws have a cancellous thread profile
3. minimal medial angulation of distal fragment
4. retroversion of the femoral head and neck (arrowhead)

Postoperative study (day 15):

1. K-E apparatus remains as before
2. early callus formation
3. note persistent dense cortical fragments
4. apposition and alignment of fragments as before

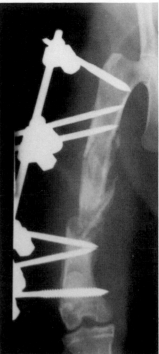

Postoperative study (day 96):

1. K-E apparatus remains, with removal of fourth and fifth fixation pins
2. minimal lucency around sixth fixation pin, indicating slight motion
3. heavy callus formation
4. dense cortical fragments have been partially incorporated in callus
5. fracture lines not visible
6. apposition and alignment of fragments as before

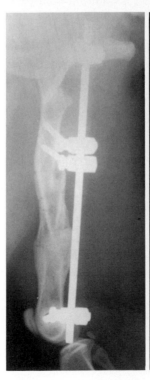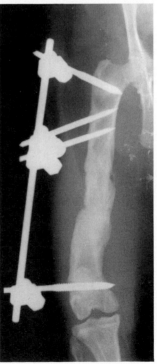

Comments on radiographic findings:

1. this case demonstrates the time required to heal a high-energy fracture—the cortical fragments still retained part of their original density and shape on the last study, more than 3 months from the time of injury; this means that there is still a delay in development of the extraosseous blood supply
2. lucency around the distal pin as a result of motion should be noted as an indicator of problems ahead
3. cat was ambulatory at the time of the last study, but removal of the remaining fixation pins was delayed for another month

Summary:

This is a serious injury, treated with an external K-E device. As the major fragments are extended into a more normal position, the surrounding muscle forms a compacting envelope that presses the midshaft fragments into better position. Use of a cancellous graft was appropriate. The pin tips are well placed with adequate insertion through the far cortex, but without protrusion into the muscle. The number of pins is appropriate and maximizes the rigidity. They are correctly angled. The resulting slight varus deformity is acceptable and results in forcing the femoral head further into the acetabulum. There is, however, a 25° anteversion of the femoral head and neck that cats adapt to fairly readily. Most small patients accept external fixation without difficulty. As healing progresses, some of the fixation pins can be removed to help avoid stress protection.

A fracture of this type could not be handled as well with an IM pin and cerclage wires. A bone plate would be more expensive and require protracted surgery because of the multiple fragments and the requirement for the plate to extend the length of the bone.

Use of the K-E device resulted in a longer healing time and less-than-perfect anatomic reduction, but it is less damaging to the soft tissues and is less expensive. The K-E device is an appropriate choice in this patient. The use of threaded pins ensured the stability required for good healing.

Signalment: 6 months old, female, intact, domestic long-haired cat

History: hit by a car

Physical Examination: right limb lameness and pain

Radiographic Examination: right femur

Study at time of entry (single lateral view) (day 1):

1. midshaft oblique fracture
2. overriding of the fragments
3. single bone fragment in soft tissues medially
4. hip and stifle joints appear radiographically normal
5. physeal plates typical for age of kitten

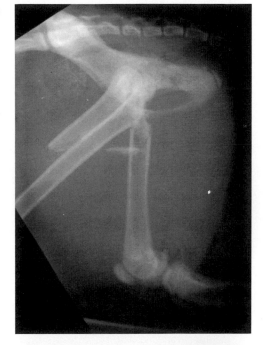

Postoperative study (single lateral view) (day 2):

1. placement of a single, large, smooth IM pin
2. bayoneting of fragments at fracture site
3. marked retroversion of femoral head
4. soft tissue bone fragment as before
5. wire skin sutures

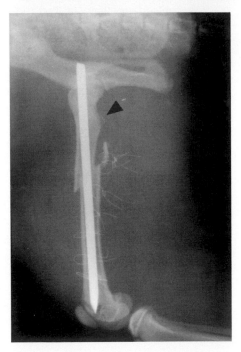

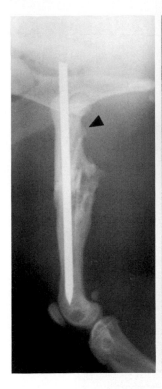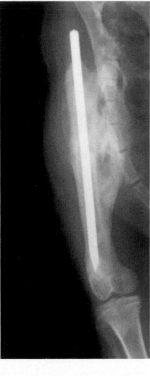

Postoperative study (day 45):

1. IM pin as before with tip in lateral condyle
2. heavy callus formation with disappearance of the fracture line
3. callus formation surrounding isolated bone fragment (arrow)
4. closure of distal femoral physeal plate
5. 2 mm femoral shortening

Comments on radiographic findings:

1. note how the lesser trochanter can be visualized on the lateral view when the femoral head and neck are retroverted (arrowheads)
2. appearance of the fracture site on the last study is typical for fracture healing interrupted by collapse around the IM pin; the heavy cortical fragments have not been resorbed, indicating avascularity
3. the large "bucket handle" callus is a result of torn periosteum, common in an immature patient
4. bone length increases by physeal growth that continues during fracture healing; however, collapse at the fracture site shortened the bone, resulting in a 2-mm loss of length.

Summary:

With a short proximal fragment, it is possible to introduce the IM pin in a normograde direction, starting the tip at the greater trochanter. This placement technique in the cat is advantageous. The interlocking of fragments is frequently used to control rotation. It is not difficult to palpate the greater trochanter during fracture reduction to identify correct reduction of the proximal femoral fragment. However, with comminution, it is more difficult to ascertain the correct alignment.

In this case, adequate control was not provided to prevent collapse around the pin and malalignment of the fragments, as indicated by the retroversion and shortening.

The fracture proceeded to heal with an exuberant callus, the result of motion, presence of a large avascular fragment, and periosteal stripping in an immature patient. The surgeon needs to understand the causes for unusual callus formation to avoid misinterpretation and possible unnecessary treatment.

Signalment: 8 months old, male, intact, domestic short-haired cat

History: cat found by the side of the road yesterday

Physical Examination: suspected fracture of right femur

Radiographic Examination: right femur

Study at time of entry (day 1):

1. distal physeal fracture (Salter-Harris type II)
2. minimal displacement of fragments
3. radiographically normal hip and stifle joint
4. growth plates as expected in a kitten of this age

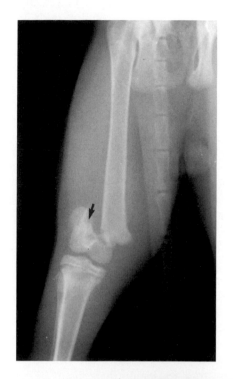

Postoperative study (day 3):

1. reduction using a single IM pin
2. excellent apposition and alignment of fragments
3. minimal soft tissue gas

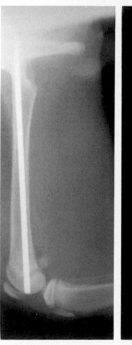 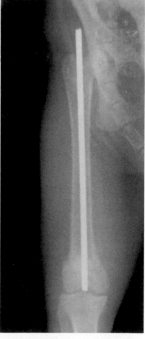

Comments on radiographic findings:

1. distal physeal fractures in the immature dog and cat usually are type II and are characterized by a metaphyseal fragment (arrow)
2. the displaced patella will be reduced at the time of fracture stabilization
3. note that the pin on the lateral view appears to be seated in the subchondral bone, whereas on the craniocaudal view the tip extends into the intercondylar space and is seated distal to the trochlear surface

Summary:

This is a common type II physeal fracture with a prominent triangular metaphyseal fragment. The surgeon chose to use a rather large IM pin that fills the marrow cavity at its narrowest point. At open reduction, the surgeon has the choice of introducing the pin at the intercondylar space, reducing the fracture, and driving the pin up the shaft of the femur to exit at the trochlear fossa; or introducing the pin into the proximal fragment, reducing the fracture, and driving the pin into the distal fragment. The distal tip seats nicely within the epiphysis, where it is within the intercondylar area. The reduced fragments are held together because of strong muscle spasms. Because of the interlocking nature of the physeal–metaphyseal interface, the pin needs only to control the fragment. The patella was easily relocated at the time of fracture reduction.

The use of the IM pin is especially appropriate for the cat because of the absence of bowing of the femur, which permits the pin to seat distally within the epiphysis. In the dog, the femur bows, making the distal seating of the pin less stable. Stabilization of the epiphysis in the dog is assisted by using a single pin inserted from the intercondylar area proximally to penetrate the cranial cortex of the proximal fragment. Longitudinal growth will continue after placement of a smooth IM pin that is perpendicular to the physeal plate.

The radiographic appearance of the healing pattern around a fracture of this type is not well described. If the growth plate is to remain open and permit continued growth of the femur, there should be no bridging callus. So, in the absence of callus, what radiographic pattern of healing is seen? It is hoped that the metaphyseal fragment will heal with bony callus, although with anatomic reduction, the callus will be small and the physeal injury will heal with reinstitution of the cartilage growth plate without any external callus being formed. Healing of physeal fractures is best evaluated by considering the clinical status of the patient rather than the radiographic evidence.

Signalment: 8 months old, female, intact, mixed breed Australian Shepherd Dog

History: dog seen to be hit by a car

Physical Examination: crepitus in left femur

Radiographic Examination: left femur

Study at time of entry (single lateral view) (day 1):

1. transverse distal metaphyseal fracture, Salter-Harris type II
2. minimal displacement of fracture fragments
3. radiographically normal hip and stifle joint
4. growth plates as expected for a puppy 6 to 8 months of age

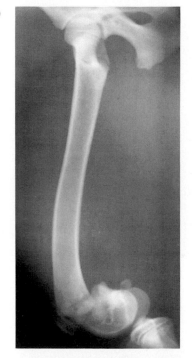

Postoperative study (day 2):

1. reduction using a single, small, smooth IM pin
2. apposition and alignment of fragments near-anatomic
3. minimal postsurgical soft tissue gas, with pocketing

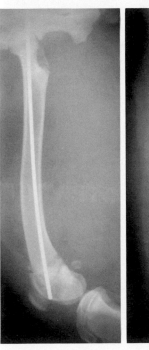
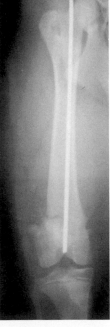

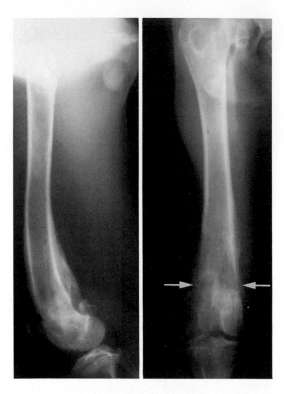

Postoperative study (day 35):

1. pin is removed
2. fracture line not visible
3. apposition and alignment of fragments are good
4. note the proximal shift of the fracture site (arrow)
5. apparent closure of the physeal plate

Comments on radiographic findings:

1. determination of whether the fracture enters the physeal plate is difficult; even if the fracture avoids the growth plate, the possibility of the trauma affecting bone growth should be considered, and owner should be warned of the possibility of growth abnormality (this probably is a type II physeal fracture)
2. injury to the femoropatellar joint must be considered in a fracture of this nature
3. stripping of the periosteum from the distal end of the proximal fragment caudally has created a void that the subperiosteal callus fills (arrows)
4. note the proximal location of the nutrient foramen

Summary:

Selection of a smooth, flexible pin seated into the distal epiphysis was appropriate. Strong muscle pull plus the irregular surface of the physeal plate ensures that the fragments are held together firmly after anatomic position is recreated.

The closure of the growth plate in a patient of this age only slightly influences the length of the femur, and will not be of clinical significance.

Signalment: 2 years old, male, intact, domestic short-haired cat

History: found to be acutely lame in left pelvic limb

Physical Examination: femoral fracture

Radiographic Examination: left femur

Study at time of entry (day 2):

1. transverse femoral fracture at the junction of middle and distal thirds with minimal comminution
2. large butterfly fragment from the craniomedial cortex of the proximal fragment
3. overriding of fragments
4. hip and stifle joints appear radiographically normal

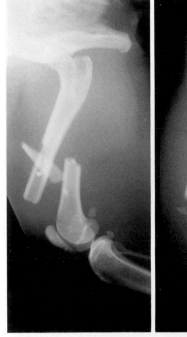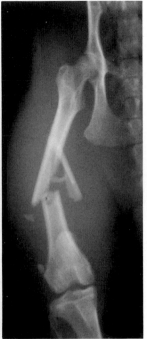

First postoperative study (single lateral view) (day 2):

1. placement of a type I external K-E apparatus with one fixation pin proximally and one pin distally (1/1)
2. proximal fixation pin does not contact far cortex
3. placement of an IM pin with distal tip in lateral condyle
4. use of two cerclage wires to reposition the butterfly fragment

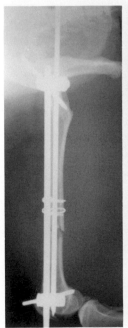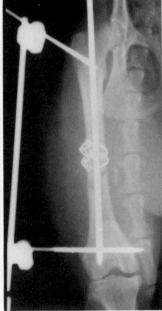

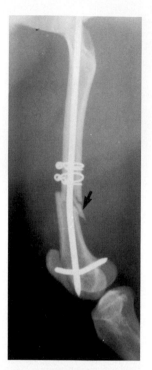

Postoperative study (single lateral view after cat struggled during recovery) (day 3):

1. proximal fixation pin removed
2. bending of distal fixation pin
3. caudal bending of IM pin
4. fragmentation of caudal cortex at fracture site (arrow)
5. caudal angulation of distal fragment

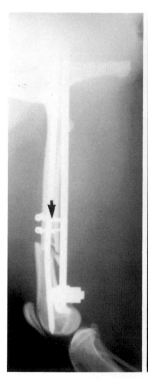
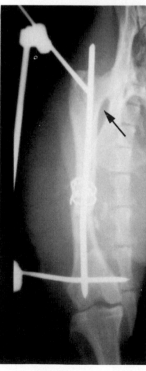

Second postoperative study (day 3):

1. replacement of proximal fixation pin without contact with the far cortex
2. distal fixation pin remains bent as before
3. IM pin remains bent
4. further fragmentation of caudal cortex at fracture site with linear fracture line extending into the proximal fragment (arrow)
5. caudal and medial angulation of distal fragment

Postoperative study (day 64):

1. IM pin remains
2. proximal fixation pin has broken through cortex (arrow)
3. cerclage wires remain
4. distal fixation pin remains (bent as before)
5. healing with callus formation
6. impaction (5 mm) at fracture site
7. fracture lines still partially identified
8. angulation of distal fragment as before
9. retroversion of the femoral head and neck

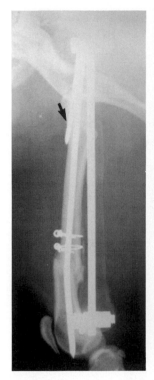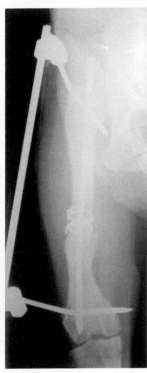

Comments on radiographic findings:

1. this case is unusual because of the abuse given the external K-E device by the patient; the full extent of additional injury to the fracture site as indicated by fragmentation and cortical splitting was not noted until the second postreduction study

Summary:

A typical problem for fracture treatment is presented by a transverse fracture and a large butterfly fragment. The first reduction was adequate, except that the proximal fixation pin did not engage the far cortex. The cerclage wires held the butterfly fragment in position. The IM pin was small, and should have been large enough to fill the medullary cavity.

The K-E apparatus became caught in the cage during recovery and the cat struggled. Because of the loose attachment of the proximal fixation pin, it became disengaged. It is interesting to note that the cerclage wires were strong enough to hold. In addition, the IM pin bent at the fracture site.

The proximal fixation pin was replaced with the tip placed cranial to the IM pin. Again, the tip failed to penetrate the far cortex. A greater space is available for proximal fixation pins caudal to the IM pin (arrow). On the last study it can be seen how this fixation pin broke through the cranial cortex.

In spite of these difficulties, the fracture healed with some shortening, a slight medial and caudal angulation of the distal fragment, and retroversion of the femoral head and neck. This case again demonstrates the importance of patient management. In many instances a very well performed surgery is invalidated by postoperative inattention. The clinician must always expect the unexpected.

CASE 14 Signalment: 3 years old, male, intact, mixed breed shepherd dog

History: dog injured left pelvic limb 13 months ago; proximal femoral fracture was stabilized by use of a bone plate, multiple cortical and cancellous bone screws, one cerclage wire, two hemicerclage wires, one with a small cross pin; healing of the fracture was noted after 8 months

Physical Examination: patient to be examined for removal of orthopedic devices

Radiographic Examination: left femur

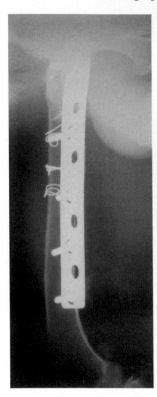 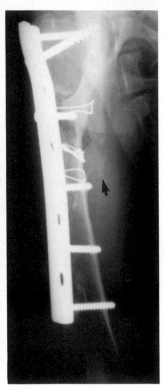

Postoperative study (to evaluate for plate removal) (13 months):

1. apparent fracture healing
2. plate with only six screws (note second screw is cancellous)
3. two hemicerclage wires in place, one cerclage wire appears broken
4. good apposition and alignment of fragments
5. marked posttraumatic modeling of femoral neck
6. bizarre callus formation on medial aspect of fracture site (arrow)
7. stress protection with thinning of the cranial cortex
8. probable generalized osteopenia due to disuse

Postoperative study (13 months):

1. removal of bone plate and screws
2. cerclage and hemicerclage wires and small K wire remain in place
3. good apposition and alignment of fragments
4. coxa vara deformity of femoral neck with retroversion
5. marked posttraumatic modeling of femoral neck as before
6. fracture line poorly evaluated
7. absence of lateral cortex just distal to the greater trochanter (arrow)
8. extension of callus into soft tissues (arrow)
9. note radiolucent screw holes
10. wire skin sutures

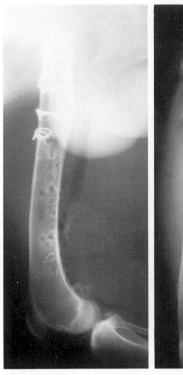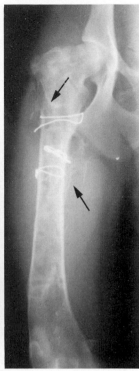

History: plate removed 13 months after original injury—patient refractured femur while walking down stairs 7 months later

Postoperative study (20 months):

1. oblique pathologic fracture in proximal half of the previously fractured femur
2. minimal displacement of the fragments
3. cerclage and hemicerclage wires and K wire remain
4. filling of old screw holes
5. soft tissue ossification secondary to original healing
6. modeling of femoral neck as before with varus deformity and retroversion

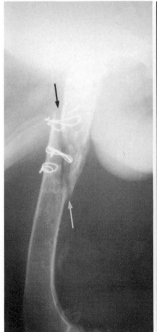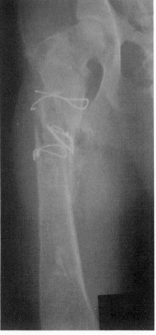

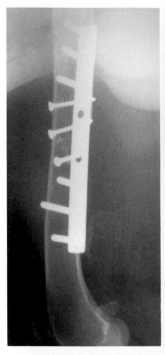
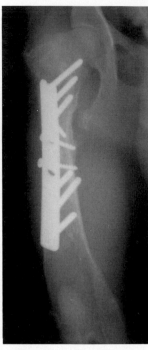

Postoperative study (30 months):

1. bone plate with three interfragmentary screws used to stabilize the pathologic fracture
2. healing with fracture line not identified
3. modeling of external callus
4. cerclage and hemicerclage wires removed
5. apposition and alignment of fragments are good
6. modeling of femoral neck with varus deformity and retroversion as before
7. soft tissue calcification

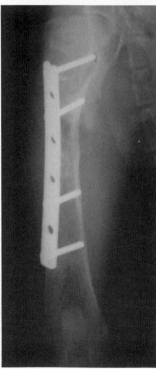

Postoperative study (to evaluate for removal of second plate) (34 months):

1. plate remains in position with removal of two plate screws and the three interfragmentary screws
2. fracture healing
3. good fragment apposition and alignment achieved
4. persistent soft tissue calcification at the fracture site and distally
5. persistent retroversion of the femoral head with posttraumatic arthrosis

Comments on radiographic findings:

1. the original study is compromised because the proximal femoral fracture is su-perimposed over the ischium and the character of the healing fracture is difficult to evaluate; a lateral view of the femoral head and a frog-leg view (hyperflexed) of the femur might have been included
2. a study of the early radiographs shows only minimal evidence of bony modeling, which may be posttraumatic or may be due to hip dysplasia—the modeling of the femoral neck noted at 13 months is thought to be secondary to a fracture line that extended into the trochanteric region, in addition to the changes secondary to the hip dysplasia
3. marked tunneling within the caudal cortex was noted at the time of healing of the original fracture

Summary:

This patient was presented for removal of a bone plate in what was assumed to be a healed fracture. The radiographic study was not complete and the region of the earlier fracture was not well visualized. The plate was removed 13 months after the original fracture. There was stress protection due to use of the plate. After removal of the plate, the expected modeling of bone tissue that usually follows fracture repair did not occur. The marked modeling of the femoral neck may be partially related to earlier hip dysplasia and secondary arthrosis, or the original fracture may have involved the proximal portion of the femur.

The dog refractured the femur 7 months later when walking downstairs. It is assumed that the new fracture was caused by failure of modeling at the original fracture site, leaving the bone weakened. Note that the refracture occurred near the site of the original fracture. This fracture was adequately treated by use of another bone plate. Note the excellent use of the interfragmentary lag screw technique. Excellent healing with good apposition and alignment of the fragments was obtained, except for persistent retroversion of the femoral head and neck.

Screws from the second plate were removed in the first stage of plate removal in an effort to correct the stress protection that was so important earlier. It is remarkable that the surgeon removed the interfragmentary screws, which ordinarily are not associated with stress protection. Dynamism, permitting stages of increased stress through partial removal of screws, does not pose as great a risk of refracture. This type of hardware removal provides the body an opportunity for bone modeling that restores the bone's original strength over a period of time. Although there is some question as to the value of this procedure, it seems to be warranted considering the early experience in this patient.

It might have been more appropriate for the primary surgeon to have made use of interfragmentary lag screws in addition to placement of the bone plate. The reliance on cerclage wires, in as large and active a dog as this one, may have been misguided.

CASE 15
Signalment: 1 year old, female, intact, German Shepherd Dog

History: referral radiograph revealed a femoral fracture

Physical Examination: marked soft tissue swelling with crepitus

Radiographic Examination: left femur (pelvis)

Referral radiograph at time of entry (ventrodorsal view of pelvis) (day 1):

1. segmental femoral fracture at junction of the proximal and middle thirds, with splitting of the segmental fragment
2. large butterfly segment from distal fragment
3. marked displacement of fracture fragments
4. radiographically normal hip joint
5. pubic fractures (arrow)
6. luxation opposite femoral head

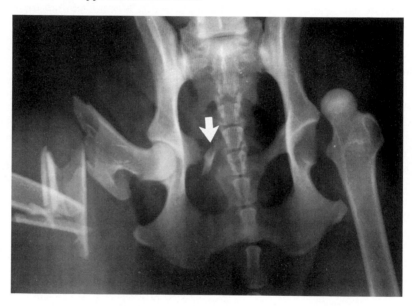

Postoperative study (day 1):

1. reduction using a contoured bone plate, three cerclage wires, and five interfragmentary screws
2. two proximal screws are cancellous—third screw has a nut and washer in position
3. anatomic apposition and alignment of fragments
4. position of femoral head and neck not well evaluated
5. marked postsurgical soft tissue gas, with pocketing

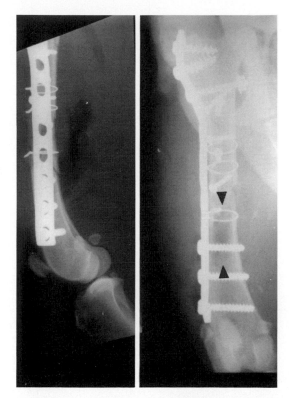

Postoperative study (day 36):

1. plate is bent at level of bone midshaft (arrow)
2. caudal and medial angulation of the distal fragment
3. early callus formation trying to bridge the fracture site
4. fracture line remains visible
5. avascular fragment at fracture site without bony response
6. retroversion of femoral head and neck

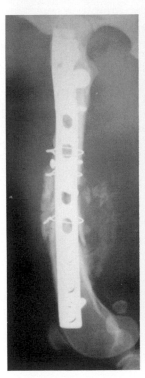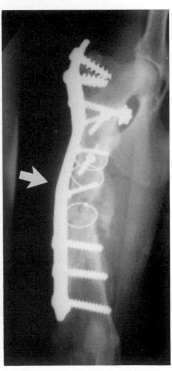

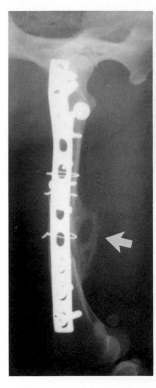
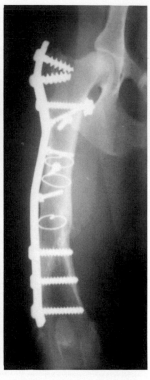

Postoperative study (2 years):

1. metallic orthopedic devices as before, with no further bending of the plate
2. healed fracture with minimal medial angulation of distal fragment
3. "bucket handle" callus formed caudally (arrow)
4. retroversion of the femoral head as before
5. healed pubic fractures

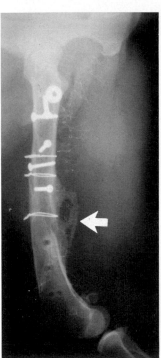
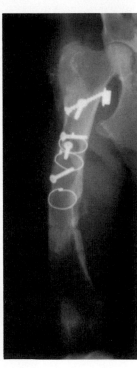

Second Postoperative study (2 years):

1. fracture healed with minimal medial angulation of distal fragment
2. plate and screws removed
3. cerclage wires, interfragmentary screws, and single nut and washer remain
4. "bucket handle" callus as before (arrow)
5. retroversion of the femoral head
6. note the screw holes
7. wire skin sutures

Comments on radiographic findings:

1. the nut and washer used on one of the proximal screws indicates that the far cortex was overdrilled, and they were required for holding the screw in place
2. lucency at the fracture site on the immediate postoperative study is due to the superimposition of soft tissue gas (arrowheads)
3. early callus formed without a pattern and was highly suggestive of infection at the fracture site; however, the dog was using the limb and lacked any clinical signs of osteomyelitis
4. note the failure to complete modeling even after 2 years—in a young patient, modeling usually occurs quickly, so that it becomes difficult to identify the original fracture—in this patient, modeling was delayed possibly because of stress protection by the plate—there is a question as to whether modeling will occur after removal of the plate
5. note the development of the "bucket handle" callus (arrows)

Summary:

This proximal femoral fracture presents a number of difficulties due to the split segmental fragment, the short proximal fragment, and the large butterfly fragment. Interfragmentary lag screws plus three cerclage wires restored the shape of the segmental fragment and held it in position. The surgeon chose an adequately long plate and accurately contoured it, using it as a neutralization plate. Cancellous screws were chosen for proximal placement because the dog is young (this breed frequently has relatively soft bone).

Healing was delayed and callus formed in an uncertain manner, without a solid bridge. Minimal medial angulation of the distal fragment indicates cortical weakness with resulting bending of the plate medially at this site. The fracture eventually healed and the dog became asymptomatic.

The possibility of metallic devices causing pain, or of becoming a site of infection or of a possible malignant transformation was considered principally because of the dog's young age. Therefore, the plate was removed, leaving the screws and cerclage wires behind. These may be left in without risk of deleterious sequelae. Immediately after the removal of the plate some form of splinting, such as a Robert Jones splint, needs to be applied, and the patient's activities curtailed.

CASE 16 **Signalment:** 2 years old, male, intact, Labrador Retriever

History: patient struck by a car earlier this day and is lame; had earlier stifle joint surgery to reduce fracture of the lateral trochlea

Physical Examination: swollen left pelvic limb

Radiographic Examination: left femur

Study at time of entry (day 1):

1. transverse fracture of midshaft femur
2. minimal comminution with small butterfly fragment
3. linear fracture line extending into the proximal fragment (arrow)
4. marked overriding of fragments
5. radiographically normal hip joint
6. chronic postoperative and probably posttraumatic changes in stifle joint

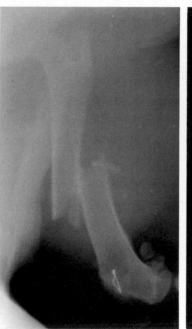 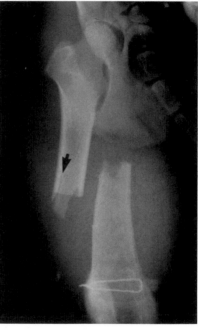

Postoperative study (day 4):

1. placement of single smooth IM pin
2. apposition and alignment of fragments good, with some distraction at the fracture site
3. probable cortical fragment laterally at fracture site (arrows)
4. postsurgical soft tissue gas
5. wire skin sutures
6. stifle joint as before

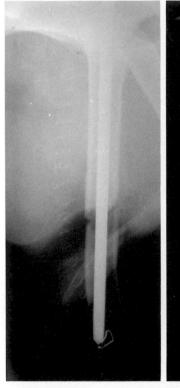
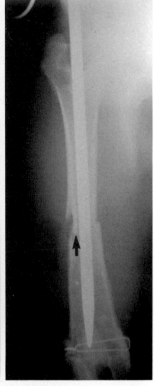

Postoperative study (day 33):

1. IM pin remains as before
2. external callus formation trying to bridge the fracture site
3. absence of bony response at distal end of proximal fragment (arrows)
4. fracture line remains visible
5. approximately 4 cm shortening
6. periosteal response proximally and distally

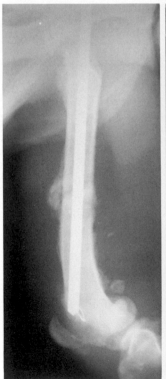
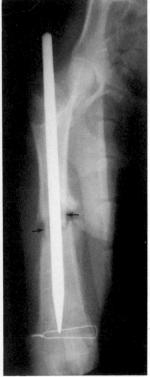

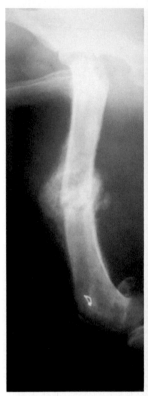
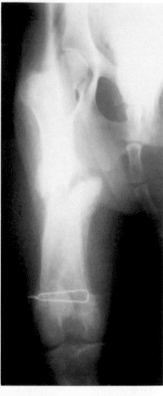

Postoperative study (day 60):

1. removal of the IM pin
2. instability at the fracture site
3. callus formation but no bridging of the fracture site
4. fracture line still visible
5. cortical lucency (osteomyelitis)
6. mature smooth periosteal new bone involving most of the shaft of the bone (osteomyelitis)
7. anteversion of the femoral head

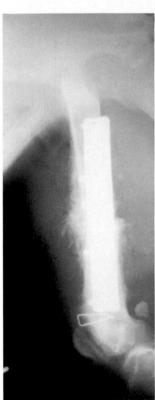
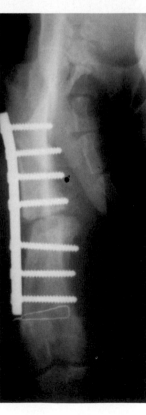

Second surgery study (day 65):

1. bone plate positioned with seven screws, four proximally and three distally
2. fracture site freshened surgically with an additional loss in length of 3 to 4 cm
3. cancellous graft difficult to separate from the earlier bone response
4. good apposition and alignment of the fragments, with minimal lateral and cranial angulation of the distal fragment
5. postsurgical soft tissue gas
6. wire skin sutures

Comments on radiographic findings:

1. at 1 month postreduction, radiographic changes are typical of osteomyelitis in the presence of a healing fracture—the distal portion of the proximal fragment is dead with no change in density and no periosteal formation; viable bone away from the fracture site is producing callus; periosteal response proximally and distally can be attributed to the trauma that caused the fracture, to the surgical trauma, or to bone infection; the clinical picture is important in confirming radiographic findings of osteomyelitis—the patient was febrile, not bearing weight on the limb, and obviously bothered
2. note the length of the healing bone—in this patient, 4 cm of shortening occurred, with the fragments impacting around the IM pin
3. note the importance of evaluating both views of the healing fracture—at day 60, the reaction around the fracture as seen on the lateral view might be interpreted as a healing callus; however, the craniocaudal view clearly shows the nonunion
4. the full extent of the injury to the femoropatellar joint is difficult to ascertain on these radiographs; it also is difficult to determine which changes were due to subsequent bone infection
5. the femorotibial joint appears radiographically normal, but this is better judged clinically

Summary:

The surgeon was faced with a transverse midshaft femoral fracture with some fragmentation. It is unusual to remove the small bony fragments from the fracture site, because this creates a void that must be filled with callus. Use of a bone graft assists in this filling. Choice of an IM pin is acceptable, except that its distal placement should have been parallel to the caudal femoral cortex instead of in the more cranial position. The caudal placement permits a deeper and more firm seating. Early formation of the external callus is away from the fracture site and is caused by motion and early infection. When a nonunion situation finally was recognized, the fracture fragments were freshened and a plate was applied. The slight open space between the plate and the distal fragment is probably filled with fibrocartilaginous callus, and it probably appeared to the surgeon that the plate was firmly in place against the cortex.

The original fracture would have been ideal for plating. This would have provided the fracture reduction and fixation essential to healing in a large, active, athletic dog. Plating provided control of fragment rotation, fragment impaction, and axial stability. The development of severe soft tissue injury provided the environment in which infection of the bone developed.

Three years after fracture healing, drainage appeared from a sinus track just proximal to the stifle joint. A fistulogram could have been used to locate the origin of the track. It is becoming apparent that removal of plates and screws must be considered in the event of development of pain, lameness, or draining track.

CASE 17 Signalment: 8 years old, male, intact, Dachshund

History: dog injured right pelvic limb when jumping from a couch to the floor

Physical Examination: fractured right femur

Radiographic Examination: right femur

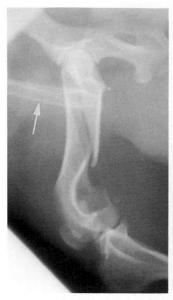
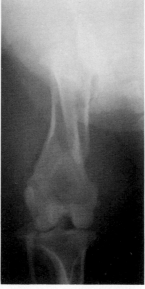

Study at time of entry (day 1):

1. long oblique fracture within the proximal half of the femur
2. minimal displacement of the fragments
3. femoral head remains within the acetabulum
4. stifle joint normal for breed

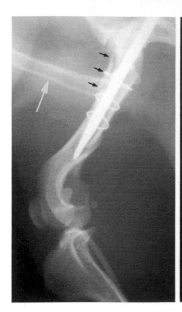
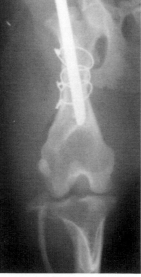

First postoperative study (day 1):

1. reduction of fracture using a single smooth IM pin and three cerclage wires
2. cranial angulation of the distal fragment
3. engagement of the distal fragment by the IM pin is minimal
4. apposition and alignment of fragments are fair
5. 1.5 cm cortical fragment fractured from the cranial cortex of the proximal fragment during reduction (arrow)
6. retroversion of the femoral head and neck

Second postoperative study (day 14):

1. larger IM pin positioned during second surgery
2. one cerclage wire removed and bone fragments impacted with loss of 0.5 cm length
3. distal tip of pin enters the condylar region
4. apposition and alignment of fragments are improved
5. normoversion of the femoral head and neck

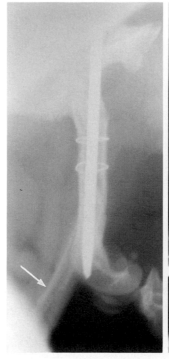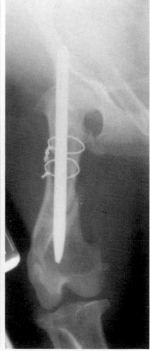

Postoperative study (day 81):

1. IM pin became loose and fell out of the limb 1 week ago, with instability at the fracture site
2. cerclage wires have moved to the fracture site
3. marked displacement of fragments
4. no callus formation noted
5. anteversion of the femoral head

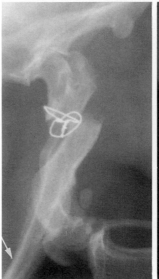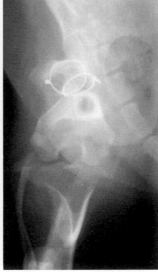

Comments on radiographic findings:

1. in a short-legged patient, it is often easier to force the pelvic limbs into flexion for ventrodorsal positioning because it is less painful—in the final study, the view of the femur is end-on, showing the medullary cavity of the distal fragment
2. in a short-legged male dog, the os penis needs to be removed from the area of interest (arrows)

Summary:

It is important to recognize the different requirements for fracture stabilization that are needed by various breeds. The chondrodystrophoid breeds have short bones with markedly deformed bone ends. This is a long, oblique low energy fracture that involves almost two thirds of the length of the femur. The IM pin was not adequately placed at the first surgery. The tip was barely inserted within the distal fragment. The three cerclage wires helped reduce the fragments and were needed to hold the cortical fragment broken during the surgical procedure. Unfortunately, the bones rotated on the pin and the fracture became unstable.

The surgeon chose to attempt a second repair using a larger IM pin that was seated further distally. Note the cranial angulation of the distal fragment that made this seating possible. Two cerclage wires were applied and the fragments bayoneted. The bones again rotated and the pin eventually fell out, leading to the nonunion as seen on the last radiograph. Note how the cerclage wires moved to the fulcrum point.

It is very difficult to plate a fracture of this type unless the surgeon uses a lag technique with the plate and provides protection for the limb through close confinement for at least 3 to 4 weeks. Perhaps in a cooperative patient it might have been feasible to use only lag screws in the treatment of the fracture, along with severe restriction of the physical activity of the patient.

What to do now after this complete failure? Suggested treatment would be the application of a small plate. It appears that two screws could be placed proximally in a lag mode and three screws distally. Fracture healing requires the dog be in cage confinement.

Signalment: 3 years old, male, intact, Golden Retriever

History: struck by a car 6 days before referral; fractured left femur had been repaired at the time of injury using a large IM pin and a half-pin K-E apparatus

Physical Examination: dog was depressed and unwilling to stand; left pelvic limb was swollen and edematous with a loose K-E apparatus in position; purulent material issued from drainage tracks

Radiographic Examination: left femur

Study at time of entry (day 7):

1. comminuted fracture at junction of proximal and middle thirds of femur
2. fracture repaired by single IM pin and half-pin K-E apparatus
3. fractured distal fixation pin
4. bone fragment in soft tissues lateral to greater trochanter (arrow)
5. hip and stifle joints appear radiographically normal

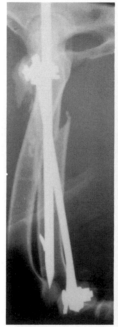
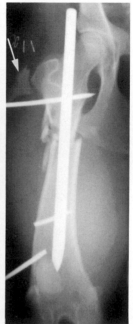

Second postoperative study (day 23):

1. removal of original IM pin and K-E device
2. placement of a bone plate with three cancellous screws proximally and four cortical screws distally
3. placement of four interfragmentary screws
4. use of large cancellous graft
5. threaded tip of distal fixation pin remains in distal fragment
6. removal of soft tissue bone fragment

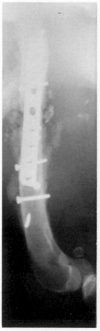
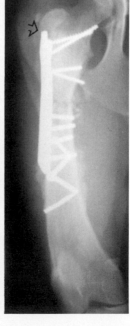

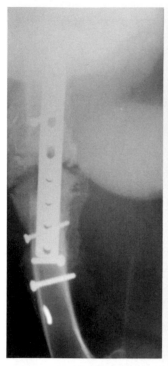
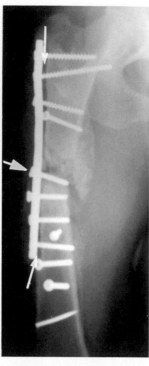

Postoperative study (day 81):

1. loosening of plate and screws from both proximal and distal fragments (arrows)
2. note the screw positioned at the fracture site (arrow)
3. callus formation surrounding fracture site except at plate location
4. fracture line widened (delayed union?)
5. distal tip of fixation pin remains in distal fragment

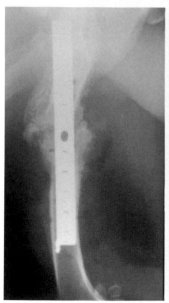
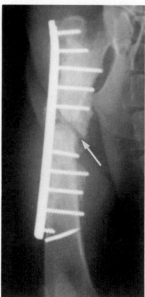

Third postoperative study (day 82):

1. removal of first bone plate and all screws
2. placement of longer plate with five screws proximally and five screws distally
3. fracture line still evident (arrow)
4. caudal angulation of distal fragment with retroversion of femoral head
5. minimal medial angulation of distal fragment
6. postsurgical soft tissue gas
7. distal tip of fixation pin remains in distal fragment

Postoperative study (day 95):

1. fragments remain stabilized with position as before
2. callus bridges the fracture line
3. healing fracture with malunion

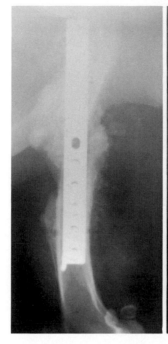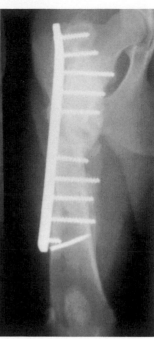

Comments on radiographic findings:

1. note the confusion in evaluation of the original lateral view caused by the superimposition of both femurs
2. with a callus formation such as seen in this patient, it is difficult to evaluate the presence of bone infection; in this case, it is assumed that the cause of the delayed union and then nonunion was motion at the fracture site, and not osteomyelitis

Summary:
The fracture originally was treated with an IM pin and a half-pin K-E device with one fixation pin proximally and one fixation pin distally. The IM pin should have followed the caudal cortex of the distal fragment so as to enter the distal femoral condyle more centrally. This requires some cranial angulation of the distal fragment. The surgeon used a bayoneting technique to interlock the fragments. This commonly used procedure resulted in a 2- to 3-cm shortening of the femur.

Motion at the fracture site was sufficient to cause breakage of the distal fixation pin at the junction of the threaded portion with the shaft. This is the weakest point in a threaded fixation pin. By noting the displacement of the two fragments of the distal fixation pin, the amount of collapse at the fracture site can be ascertained.

At referral, the fracture was plated using lag screws for interfragmentary compression. The radiographs did not show linear fractures in the distal fragment; however, they were noted at the time of surgery and required interfragmentary compression. The plate should have extended over the bone in which interfragmentary screws are placed. The selected plate was too short. The proximal screws did not seat well into the femoral neck. The plate was fitted into a slot cut into the greater trochanter (arrow). Number 5 screw placed at the fracture site interfered with healing and contributed to instability and nonunion between the major fragments.

At the next surgery, the plate was removed along with all of the screws. The fixation pin tip remained. A longer and heavier plate was used with five screws proximally and five screws distally. Note that this plate was more correctly contoured. The massive callus formed around the proximal fragment prevented the surgeon from accurately determining the shape of the original shaft, and the plate was angled relative to the fragment. In spite of all the difficulties and additional surgeries, healing was noted on the last study.

Basically, this is a difficult fracture with a short proximal fragment. The location of this fracture results in the distal fragment being a large lever arm. The comminution means that there is no clear fracture line with fragments that fit anatomically, resulting in the surgeon either accepting shortening due to collapse at the fracture site, or trying to maintain separation of the fragments by the use of a buttress plate and cancellous grafting. In the first surgery, the external fixator was entirely too small for this size of dog, and broke within several days. A decision to use a plate in a buttress mode with bone grafting to fill the gap at the fracture site was correct. However, the plate was too small, too short, lacked proper screw placement into the femoral neck, and had a screw placed into the fracture site. Motion again prevented healing and a larger plate was required. The massive callus prevented anatomic alignment of the fragments, but healing resulted.

Signalment: 5 years old, male, intact, Australian Shepherd Dog

History: dog hit by car

Physical Examination: unable to walk on right pelvic limb

Radiographic Examination: right femur

Study at time of entry (day 1):

1. long, oblique fracture midshaft femur
2. long butterfly fragment with minimal comminution (arrow)
3. marked overriding of fracture fragments
4. radiographically normal hip and stifle joints

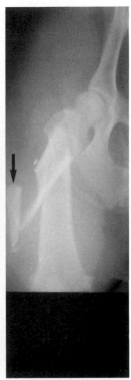

Postoperative study (day 2):

1. near anatomic reduction using a single interfragmentary screw and a nine-hole bone plate with three screws proximally and three screws distally
2. additional support achieved through a type I K-E apparatus (1/1)
3. good apposition and alignment of fragments
4. postoperative soft tissue gas

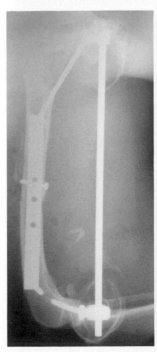
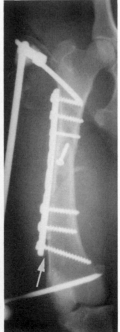

Postoperative study (day 27):

1. bone plate bent at level of fourth hole—all screws firmly attached to underlying bone
2. interfragmentary screw is loose
3. removal of K-E apparatus
4. fracture lines still identified and callus is lacking
5. butterfly fragment caudal to fracture site
6. marked caudal and medial angulation of fragments as a result of plate bending
7. drill holes for K-E apparatus still evident

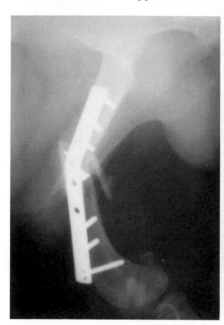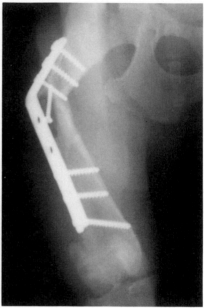

Comments on radiographic findings:

1. the bone plate was not contoured with space between the plate and the underlying bone (arrow)

Summary:

This is a common oblique midshaft fracture with a large butterfly fragment. The surgeon reduced the fracture using a single lag screw to position the butterfly fragment. A bone plate was applied with three screws proximally and three screws distally. It may be that the surgeon thought that this plate was inadequate in length and strength and applied a type I K-E device.

The technique may have been a choice between less rigid fixation and the possibility of stress protection had a larger plate been used. In a medium-sized dog, the surgeon chose the lighter plate and hoped to provide some additional support with the K-E.

The plate did not break, as it might have done with cycling from repeated motion. Instead, all of the screws held tightly to the bone and the plate bent. This indicated that the bending was the result of acute trauma. On being made aware of these facts, the owner admitted that the dog had jumped from his pickup truck, reinjuring the limb.

The bone was replated and the fracture eventually healed. The owner was much more observant of the dog's activities after paying for the second surgery.

Signalment: 8 months old, female, intact, mixed-breed dog

History: dog jumped from moving truck

Physical Examination: swelling of right pelvic limb, walks on other three limbs

Radiographic Examination: right femur

Study at time of entry (single lateral view) (day 1):

1. transverse fracture at junction of middle and distal thirds of femur
2. minimal comminution
3. marked displacement of fracture fragments with overriding
4. radiographically normal hip and stifle joint
5. growth plate closure suggesting a slightly older age than the owner reported

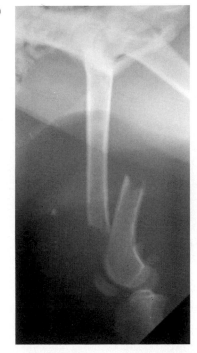

Postoperative study (single lateral view) (day 1):

1. reduction using a single large, deeply seated, threaded IM pin
2. good apposition and alignment of fragments

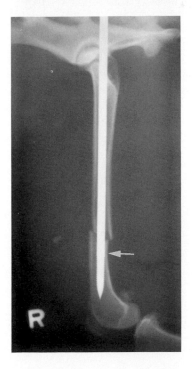

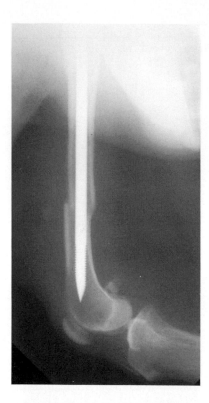

Postoperative study (single lateral view) (day 77):

1. IM pin remains in position
2. fracture line closed
3. external callus is maturing
4. apposition and alignment of fragments as before

Comments on radiographic findings:

1. with solid fixation there is no requirement for a large external bridging callus

Summary:

The surgeon chose to use a large threaded IM pin, seated deeply into the distal fragment and accurately reducing the fracture. Good stabilization was achieved even though there was a sacrifice of medullary blood supply due to the use of a large IM pin. The fragments impacted slightly so that rotation was not a problem. Note that the pin is against the caudal cortex (arrow), which directs the pin more deeply into the condylar area. The femoral head and neck are in good position.

In this patient, the surgeon chose to compromise a portion of the blood supply to the fracture site to achieve good stabilization. This often is a choice that needs to be made in deciding on the method of fixation to be used. In this case, healing with minimal callus was indicative of little motion at the fracture site. Eventually, the pin will be removed from this well treated femoral fracture.

Signalment: 13 years old, male, castrated, mixed-breed dog

History: unknown how dog injured left pelvic limb

Physical Examination: marked soft tissue swelling in pelvic limb with probable fracture of the left femur

Radiographic Examination: left femur

Study at time of entry (day 1):

1. long, oblique fracture of proximal half femur extending to the femoral neck
2. probable low femoral neck fracture (arrows)
3. one large, displaced butterfly fragment with minimal comminution
4. marked displacement of fracture fragments with overriding
5. radiographically normal hip and stifle joint

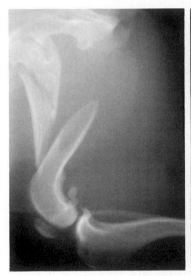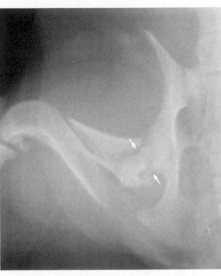

Postoperative study (day 18):

1. reduction using a single IM pin, two full cerclage wires, and an external type I K-E apparatus
2. distal tip of IM pin protrudes into the femoropatellar joint space (arrow)
3. fair apposition and alignment of fragments
4. overriding of the proximal fragments, causing shortening of the bone
5. unstable femoral head and neck fragment
6. marked varus deformity of femoral head and neck
7. small piece of wire in soft tissues

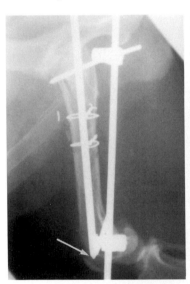
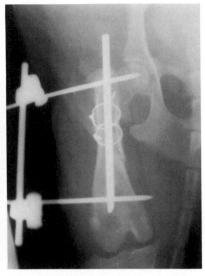

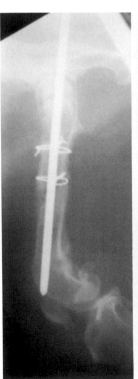
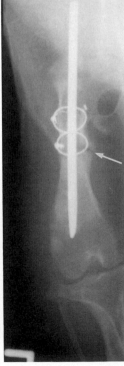

Postoperative study (day 100):

1. IM pin and cerclage wires remain in position
2. removal of external K-E apparatus
3. fracture line still visible (arrow)
4. minimal callus formation
5. apposition and alignment of fragments as before, with 1 cm shortening due to collapse at the site of the femoral neck fracture
6. retroversion of femoral head and neck

Comments on radiographic findings:

1. the appearance of the ends of the rather shortened long bones suggests that the dog's lineage included a member of a chondrodystrophoid breed
2. the evaluation of the low femoral neck fracture could have been enhanced through views made with the limb in extension
3. it is important to measure the length of the fractured bone to ascertain the amount of collapse that can occur around an IM pin

Summary:
This femoral fracture presents difficulties because of the oblique fracture line, the large butterfly fragment, the very small proximal segmental fragment, and the fractured femoral neck in a chondrodystrophoid dog. The IM pin was thoughtfully introduced through the greater trochanter and the two cerclage wires were meant to hold the butterfly fragment in position. The surgeon recognized the femoral neck fracture and used the proximal fixation pin of a type I K-E apparatus in an attempt to stabilize it. The IM pin appears to protrude through the cranial cortex of the trochlea.

Two fixation pins proximally and distally might have provided greater support. However, the surgeon must vary the technique with species and breed differences. These bones are very short and the fat thigh complicates pin placement. The fracture might have been treated by a bone plate with a proximal screw lagged across the femoral neck fracture. The butterfly fragment could have been treated with interfragmentary placement of one or preferably two screws. The plate would have extended from the trochanter as far distally as possible.

However, if the bone shortening and the danger of nonunion can be accepted, the method applied seemed to be adequate. It is unfortunate that the IM pin entered the joint. The fracture was healed 1 year after the injury. This dog had a functional limb and performed well during the remaining days of its life, even though the fracture healing included marked varus deformity at the hip and some problems in stifle joint extension.

Signalment: 2 years old, male, castrated, domestic short-haired cat

History: cat missing for 7 days and returned nonweight bearing on right pelvic limb

Physical Examination: marked swelling of right pelvic limb

Radiographic Examination: right femur

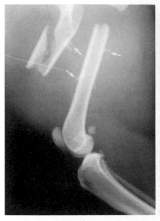 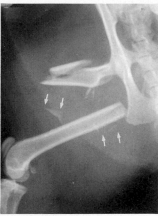

Study at time of entry (day 1):

1. transverse fracture at junction of proximal and middle thirds of the femur
2. butterfly fragment originating from the proximal fragment
3. overriding of fragments
4. debris on skin mimics bone fragmentation (arrows)
5. hip and stifle joints appear radiographically normal
6. fracture appears to be acute

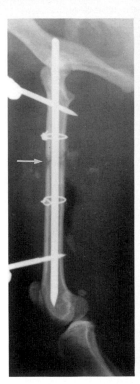 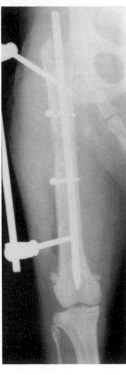

Postoperative study (day 4):

1. placement of a type I external K-E apparatus with one pin proximally and one pin distally (1/1)
2. placement of an IM pin
3. proximal cerclage wire reduces the butterfly fragment
4. distal cerclage wire reduces linear fracture line within the distal fragment detected at the time of surgery
5. note the bone loss at the fracture site (arrow)
6. apposition and alignment of fragments appear good

Postoperative study (day 64):

1. fracture line still identified (arrow)
2. minimal callus formation
3. metallic devices remain as before without any signs of loosening
4. apposition and alignment of fragments remain

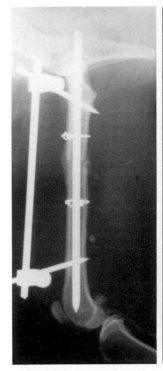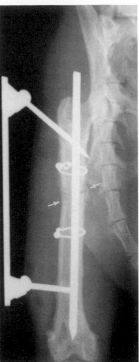

Comments on radiographic findings:

1. linear fracture lines leading from the fracture site may not be identified on the original radiographs, and examination at the time of surgery is essential

Summary:

This is a difficult fracture due to the short proximal fragment with its large butterfly fragment. The IM pin was deeply seated in the midshaft of the femur. The cerclage wires were placed to reduce the proximal butterfly fragment and to prevent further extension of a linear fracture in the distal fragment. Because of the tendency for axial movement, the K-E device was applied.

In the dog, single fixation pins proximally and distally often are not sufficient, but this technique works well in the cat. The fracture healed slowly, probably because of the assumed 7-day delay in treatment. As a result of the rigid fixation, as indicated by the absence of any radiolucency around the fixation pins plus the selection and introduction of a very adequate IM pin, this difficult fracture was managed in an exemplary manner.

CASE 23 Signalment: 2 years old, female, spayed, domestic short-haired cat

History: unknown trauma

Physical Examination: right limb lameness and pain

Radiographic Examination: right femur

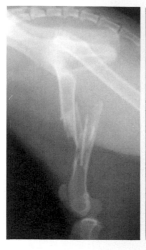 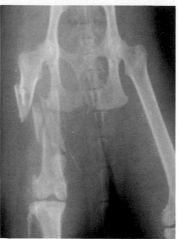

Study at time of entry (day 1):

1. midshaft spiral fracture of femur with a longitudinally split segmental fragment
2. overriding of the fragments with comminution
3. hip and stifle joints appear radiographically normal

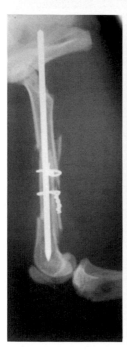 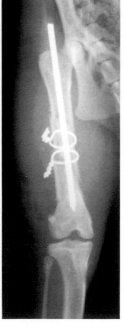

Postoperative study (day 2):

1. placement of a single, large, smooth IM pin
2. two cerclage wires used to reduce butterfly fragments
3. good apposition and alignment of fragments with minimal cranial angulation of the distal fragment
4. soft tissue gas

Postoperative study (day 43):

1. heavy callus formation distal and medial to fracture site
2. impaction at fracture site with 0.5 cm shortening of bone
3. further cranial and lateral angulation of distal fragment, indicating instability
4. heavy cortical fragments still present (arrow)
5. fracture lines still identified
6. IM pin has been removed
7. cerclage wires remain

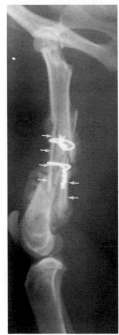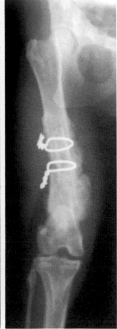

Comments on radiographic findings:

1. appearance of the fracture site on the last study is difficult to evaluate because of the effect motion had on delaying resorption of the heavy cortical fragments; lack of resorption indicates an absence of a blood supply—avascularity usually is due to motion at the fracture site, but may indicate an osteomyelitis—clinically, the cat showed no evidence of any infection—it was treated with cage rest and the fracture continued to heal, with improvement noted clinically.

Summary:

This fracture is made difficult by the split segmental fragment, the short distal fragment, and the multiple butterfly fragments. Cat bone often appears more brittle and fragments more readily than does dog bone. Fractures of this type are therefore not uncommon. A rather small-diameter IM pin was introduced and some of the fragments controlled with two cerclage wires. After the postoperative radiograph, the surgeon applied a type I K-E device to provide for further stability. Unfortunately, the cat was difficult to manage after reduction of the femoral fracture, and the K-E apparatus was torn off as the cat broke through a window screen while jumping to the street below. The IM pin also was backed out and was removed by the owner. The cat was confined and healing progressed, with formation of an exuberant callus around a malpositioned distal fragment, leading to malunion.

Theoretically, it would have been possible to use a bone plate in a neutralization mode. The next possibility would have been to use an external K-E apparatus with two fixation pins proximally and two fixation pins distally (2/2). Using this technique, it would still have been necessary to control the cat physically. The case illustrates the problems associated with the use of IM pins and cerclage wires with external support from a type I K-E device. It also shows the absolute necessity for patient control.

So, what to do? It would be very difficult to plate at this time; because of the callus formation, little would be gained. Sometimes the surgeon must accept the consequences of failure and allow for healing in a less-than-desirable manner.

Tibia

Signalment: 8 months old, male, intact, domestic short-haired cat

History: cat struck by a car

Physical Examination: hematuria—suspected fracture of right tibia

Radiographic Examination: right tibia

Study at time of entry (day 1):

1. oblique fracture at midshaft of right tibia with a single butterfly fragment (arrowheads)
2. minimal overriding of fragments
3. fibula intact with luxation of the fibular head (arrow)
4. radiographically normal stifle and tibiotarsal joints
5. physeal plates typical for a cat of this age

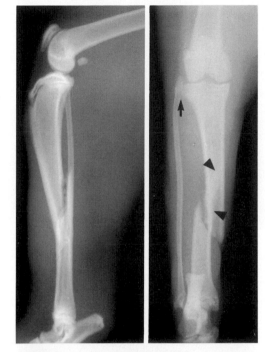

Postoperative study (day 3):

1. reduction using a single IM pin and a single cerclage wire
2. butterfly fragment not stabilized
3. minimal soft tissue gas
4. fibular head reduced but not stabilized

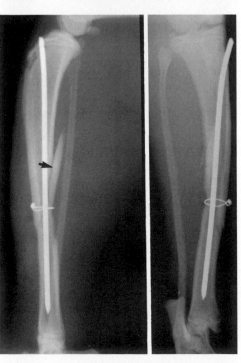

257

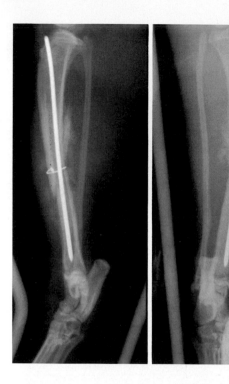

Postoperative study (day 16):

1. IM pin and cerclage wire remain in position
2. 1 cm shortening due to slippage at the fracture site
3. early fluffy external callus formation without pattern
4. caudal and lateral angulation of the distal fragment
5. reluxation of the fibular head
6. Schroeder-Thomas splint

Comments on radiographic findings:

1. luxation of the fibular head could easily be ignored
2. callus formation at 16 days is not "healthy" in appearance—it forms away from the fracture site and looks more like inflammatory periosteal response; this is suggestive of an abortive attempt at callus formation in association with movement at the fracture site

Summary:

Open reduction was used with this tibial fracture to place a small IM pin, introducing it from the craniomedial aspect of the proximal tibia. Use of a single cerclage wire usually is considered insufficient because it provides a fulcrum for fragment movement. The butterfly fragment obviously is not well positioned (arrow). Multiple wires will ensure that fragments such as this do not move. Some means to prevent slippage of the cerclage wire need to be taken, such as: (1) notching the bone, (2) using a small K wire, or (3) a hemicerclage technique.

The surgeon realized that the required degree of stability had not been obtained and decided to use a Schroeder-Thomas splint. This may cause a problem because the Schroeder-Thomas is a traction splint and the IM pin and wire depend on spasm of the muscle to draw the fragments together. Thus, one mode is counteracting the other. There also is the problem of the weight of the splint. If it is not perfectly applied, it becomes a pendulum, causing motion at the fracture site. If additional fixation were needed, a coaptation splint would have been a better choice.

Luxation of the fibular head apparently was not observed. Because it is the site of insertion of the lateral collateral ligament of the stifle joint, its surgical replacement would have been appropriate.

In a little over 2 weeks, control of the butterfly fragment was lost. The manner of callus formation suggested the possibility of an eventual nonunion.

The patient was lost to follow-up, but in a young cat it is entirely possible that the fracture healed with an exuberant callus. This fracture could have been handled in a more appropriate manner using external fixation or a small bone plate.

Signalment: 4 months old, male, intact, Border Collie

History: dog was kicked by a horse 4 days ago

Physical Examination: fractured left tibia

Radiographic Examination: left tibia

Study at time of entry (day 1):

1. almost transverse fracture at junction of proximal and middle thirds of tibia
2. minimal comminution
3. lateral and caudal angulation of the distal fragment
4. 25% end-to-end apposition of the major fragments
5. transverse fibular fracture
6. radiographically normal stifle and tibiotarsal joints
7. growth plates as expected for a puppy of this age

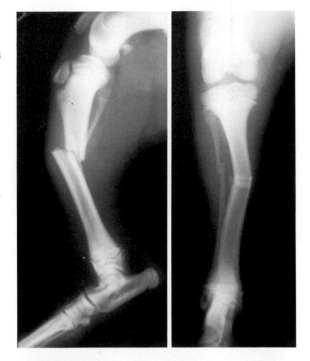

Postoperative study after splint change (day 11):

1. attempted reduction with Schroeder-Thomas splint
2. cranial and lateral angulation of distal fragment
3. 25% end-to-end apposition of fragments
4. minimal callus formation without bridging
5. disuse osteopenia in all bones

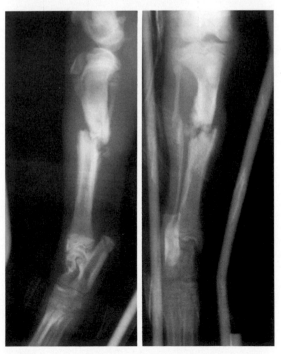

Comments on radiographic findings:

1. callus formation, even at day 11, should have bridged the fracture site in a puppy of this age—absence of callus indicates delayed healing because of motion—note the sclerosis at the fracture site, suggesting the development of a hypertrophic pattern of fracture nonunion
2. new bone response at the fracture site also may be suggestive of infection

Summary:

A seemingly straightforward midshaft transverse fracture of the tibia in a puppy led the surgeon astray. The apparently simple fracture was treated with a Schroeder-Thomas splint. This is a traction-type splint, and does not provide the stability that is regularly achieved with more current therapy. Note that the limb is in extension and the bars are improperly placed. A correctly designed splint ensures that the joints are in flexion. This is an inexpensive technique that turned out to be more expensive in the long run. The tremendous healing potential in a puppy with a low-energy fracture was countered by motion at the fracture site.

In this patient, there is evidence of development of pending nonunion. At this time, it would be appropriate to apply a bone plate, which would apply the requisite stability, along with the use of a cancellous bone graft to stimulate healing. Rongeurs would be used on the "white" ends of the bone to freshen them until blood is seen, providing vascularity to assist in the healing process.

Signalment: 6 months old, female, intact, domestic short-haired cat

History: found nonweight bearing on right pelvic limb

Physical Examination: marked swelling of right hock

Radiographic Examination: right tibia

Study at time of entry (day 1):

1. type I physeal fracture with complete displacement of the fragments
2. caudal displacement of distal fragment
3. undetermined injury to the medial or lateral malleolus
4. tibiotarsal joint difficult to evaluate radiographically
5. soft tissue swelling
6. physeal closure appropriate for reported age

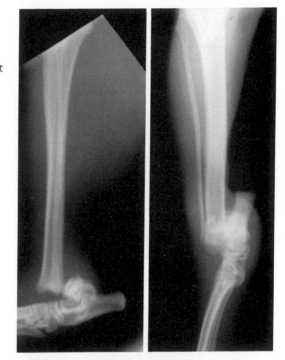

Postoperative study (day 16):

1. reduction and stabilization through use of a single, smooth IM pin crossing the physeal plate and tibiotarsal joint
2. good alignment and apposition of the fracture fragments
3. periosteal new bone on distal tibia due to periosteal stripping
4. lucency in distal tibia (arrows)

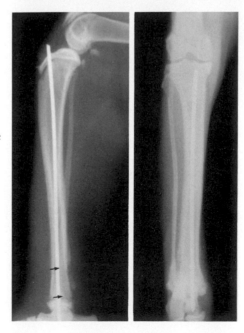

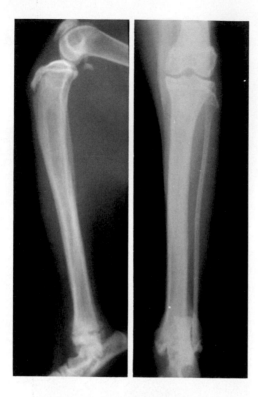

Postoperative study (day 33):

1. apposition and alignment of fragments as before
2. removal of the IM pin
3. radiolucent physeal plate
4. maturing of periosteal new bone on distal tibia

Comments on radiographic findings:

1. failure to understand possible injury to the medial and lateral malleolus is a significant problem in handling this patient
2. radiographic changes on day 16 are confusing—the apparent destructive pattern within the distal tibia is suggestive of an inflammatory process; however, the lucency could be associated with continuing bone growth
3. the widened, lucent physeal plate is as expected for a healing fracture, and represents the continued production of radiolucent cartilage
4. lack of any angulation at the physeal plate suggests that growth is continuing uniformly throughout the plate
5. radiography of the opposite limb is necessary to determine the exact effect of the trauma on bone length

Summary:

This is a physeal fracture with complete separation of the fragments. The possibility of collateral ligament tear is remote because the ligament attachment is on the epiphyses. The fracture was reduced and stabilized with the placement of a single, smooth IM pin driven in a normograde direction from the proximal tibia, crossing the affected growth plate and the tarsus, and exiting through the talus and calcaneus. The limb was then placed in a cast for additional support.

Continued bone growth can be expected because it is not interfered with by the smooth pin. The degree of damage to the articular surface is limited, and after removal of the pin clinical experience shows that permanent damage to the cartilage does not often occur. Secondary joint disease usually does not develop in cases such as this.

This treatment was well performed and resulted in good healing of a physeal fracture.

Signalment: 1 year old, male, castrated, Siamese cat

History: cat returned home carrying its left pelvic limb

Physical Examination: gunshot wound to left tibia

Radiographic Examination: left tibia

Study at time of entry (day 1):

1. comminuted spiral fracture at junction of middle and distal thirds of the tibia
2. small butterfly fragment caudally
3. marked displacement of fracture fragments with overriding
4. distal fibular fracture
5. metallic fragments within soft tissues
6. tibiotarsal joint appears normal radiographically
7. physeal plates are open, suggesting age of 8 to 10 months

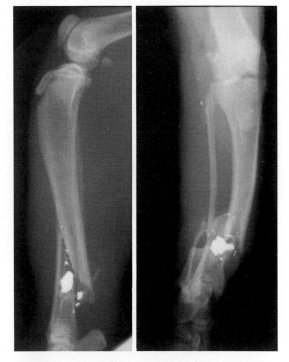

Postoperative study (day 2):

1. reduction using two cerclage wires
2. stabilization using a single IM pin with a chisel point
3. good apposition of fragments
4. slight medial angulation of distal fragment
5. large metallic fragment removed
6. fibular reduction secondary to tibial repair

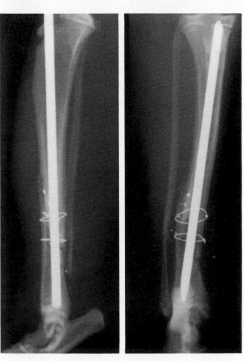

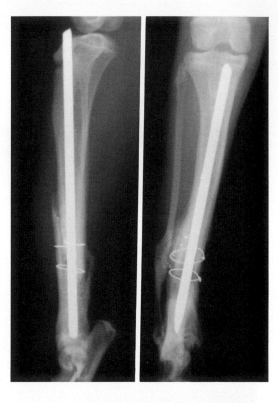

Postoperative study (day 8):

1. fracture fragments remain as before with IM pin and cerclage wires
2. heavy callus forming, with fracture lines difficult to visualize
3. synostosis forming between tibia and fibula

Comments on radiographic findings:

1. positioning for the first craniocaudal view unfortunately is not anatomic—the technician is torn between the desire to produce a radiograph with anatomic positioning and instructions not to exacerbate the injury

Summary:

This very high-energy oblique fracture is due to a gunshot wound. Treatment was done by placement of a chisel-pointed IM pin inserted normograde. The cerclage wires were small but effective. The IM pin is of good size and well seated.

This is a difficult fracture that was handled well, and healing was seen on radiographs made at the referring clinic 38 days later. The synostosis of the fibula to the tibia is not of clinical significance. Gunshot fractures often do not heal this quickly or as well. Removal of the lead fragments is rather unusual because it entails considerable further soft tissue damage.

Signalment: 1 year old, male, intact, domestic short-haired cat

History: cat returned home this morning and was painful on pelvic limbs with multiple lacerations to limbs and tail

Physical Examination: suspected fracture of right tibia

Radiographic Examination: right tibia

Study at time of entry (day 1):

1. comminuted fracture at junction of middle and distal thirds of the tibia
2. multiple fractures within the fibula, one comminuted and one with bone loss
3. caudal angulation of distal tibial fragment
4. multiple metallic fragments within soft tissue (one just caudal to the stifle joint)
5. soft tissue gas
6. radiographically normal stifle and tibiotarsal joints
7. physeal plates typical for a cat 16 to 18 months of age

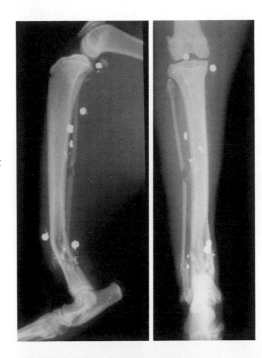

Postoperative study (day 1):

1. reduction using a type I K-E apparatus cranially with one fixation pin proximally and one fixation pin distally, and a type II K-E apparatus medio-laterally with two threaded fixation pins proximally and one threaded fixation pin distally
2. apposition and alignment of fragments good
3. metallic fragments as before

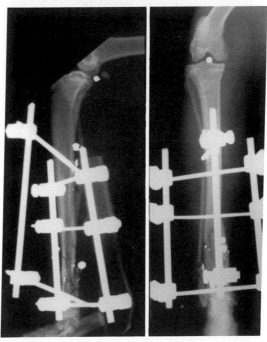

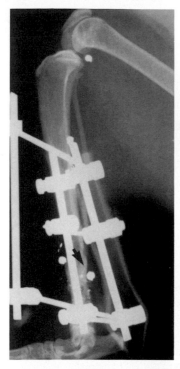
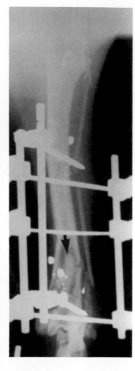

Postoperative study (day 28):

1. K-E apparatuses as before
2. fracture line poorly visualized
3. early external callus formation with some bridging of the fracture lines
4. apposition and alignment of fragments as before, with minimal medial angulation and external rotation of the distal fragment

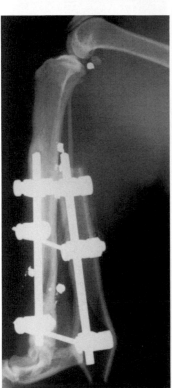
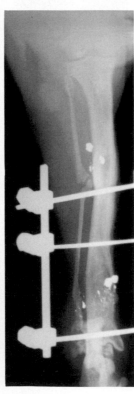

Postoperative study (day 72):

1. removal of the type I K-E apparatus—type II K-E apparatus remains as before
2. fracture line not visualized
3. callus formation bridging the fracture site
4. metallic fragments as before
5. apposition and alignment of fragments as before

Comments on radiographic findings:

1. this is a typical pattern for a shotgun injury—although the shot usually does not cause fractures in dogs, they cause much more severe injuries in the cat, as seen in this patient with multiple fractures
2. note the cortical fragment from the caudal cortex that was delayed in incorporation within the callus; this is not a sequestrum, only a dense cortical fragment that requires a longer time for incorporation (arrows)
3. note how the healing takes place, incorporating the shot into the callus
4. status of the lateral malleolus is questionable, and careful physical examination would be required to determine interference in joint function

Summary:
Because of the soft tissue injury, the choice of the K-E apparatus was a good one in this very high-energy fracture. Placement of the fixation pins was made in consideration of the weight of the fixator on the small cat, and did not include uniting the lateral bars to the cranially positioned bar. This is a variation of a type III K-E apparatus characterized by "tenting" without "triangulation." It is possible to consider treatment of this type because of the nature of the patient. This cat was a very satisfactory patient, behaving well.

This modified K-E device was the best method, considering the tissue damage and the small distal fragment. Note that the fracture healed around the metallic foreign bodies without difficulty. The metallic shot caudal to the stifle joint may attract the attention of the surgeon with its presumed need for its removal. It is better to leave it alone at this time because of the almost impossible task of locating it. In this patient, it was not causing any clinical problem.

┌─────────┐
│ **CASE 6** │ **Signalment:** 3 years old, male, castrated, domestic short-haired cat
└─────────┘ **History:** hit by car

Physical Examination: left tibial fracture, status of fibula undetermined by palpation

Radiographic Examination: left tibia

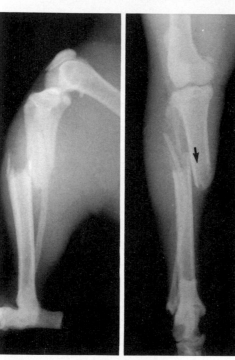

Study at time of entry (day 1):

1. slightly oblique fracture at junction of proximal and middle thirds of tibia
2. minimal comminution and possibility of fissure fracture within proximal fragment (arrow)
3. marked overriding of fracture fragments
4. proximal fibular fracture
5. gas within soft tissue
6. stifle and tibiotarsal joints appear normal radiographically

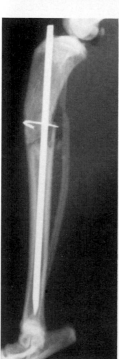
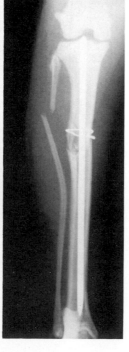

Postoperative study (day 5):

1. reduction using a single cerclage wire with stabilization achieved through use of single, smooth IM pin
2. good apposition and alignment of fracture fragments

Postoperative study (day 46):

1. unstable fracture with impaction at the fracture site
2. proximal shifting of cerclage wire 0.3 cm
3. loss of 0.5 cm bone length due to impaction
4. fracture lines remain visible
5. minimal callus formation at the fracture site

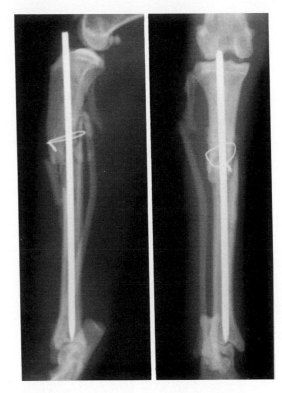

Comments on radiographic findings:

1. detection of the fissure lines on the original study is difficult, but they are more easily detected on the first postreduction study, in which new fragments are identified
2. impaction of the fragments can be appreciated by noting how the distal fragment has slipped to be included more fully by the cerclage wire

Summary:

In the reduction of this tibial fracture, the surgeon created additional fragmentation from the proximal fragment. This was poorly controlled with a single cerclage wire that did not add stability to the major fracture site. The choice of the IM pin was satisfactory, with positioning achieving contact at both cranial and caudal cortices of the distal fragment. However, within 2 weeks, there was collapse along the pin at the fracture site and a new cranial fragment broke free. After impaction, the fracture stabilized and ultimately healed.

The single cerclage wire failed in its purpose. Additional wires would have provided further support. However, a single IM pin cannot be expected to maintain length in bones with bone loss, even though small fragments have been supported by cerclage wires. Length maintenance in such fractures would require the use of an external fixator or plating.

Use of a K-E apparatus after discovery of the expected weakness at the fracture site might have helped; however, there is little space for placement of the distal fixation pin. This is an example of fracture treatment that was questionable, but with collapse around the IM pin, the fracture became more stable and eventually healed at 100 days. A minimal amount of bone shortening occurred. Healing depends on solid bone-to-bone contact which is a frequent result of a secondary collapse along a smooth IM pin.

CASE 7   **Signalment:** 3 years old, male, intact, German Shepherd Dog

History: dog was struck by a car 2 days earlier

Physical Examination: severe soft tissue injury to lower left pelvic limb

Radiographic Examination: left tibia

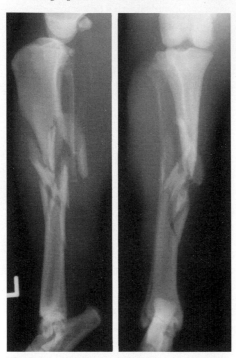

Study at time of entry (day 1):

1. highly comminuted oblique fracture of middle third of tibia
2. multiple butterfly fragments
3. minimal displacement of fracture fragments
4. radiographically normal stifle and tibiotarsal joints

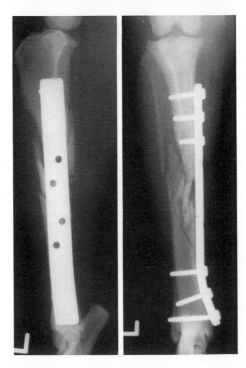

Postoperative study (day 1):

1. reduction with well contoured heavy bone plate with three screws proximally and three screws distally, used in a buttress mode
2. excellent apposition and alignment of fragments
3. postsurgical soft tissue gas

Postoperative study (day 90):

1. bone plate as before
2. fracture lines still evident, with loss of detail of fragment ends
3. minimal bridging callus
4. apposition and alignment of fragments as before

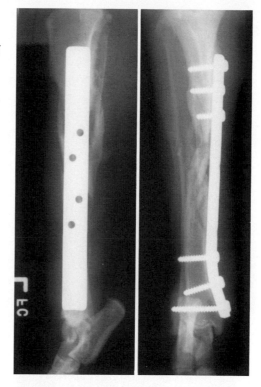

Comments on radiographic findings:

1. healing as expected around a high-energy fracture with extensive soft tissue injury; only minimal callus forming at the end of 3 months
2. note that the bone is maintaining normal density, suggesting that the limb is being used, avoiding disuse osteopenia

Summary:

The surgeon wisely did not disturb the soft tissue injury. A long bone plate was nicely contoured and placed in a buttress mode. Note that the holes in the plate are positioned so that the underlying drill holes will not be in the same plane and weaken the bone.

Reconstruction of the bone through the use of cerclage wires and interfragmentary screws might have been attempted, but would have required extensive manipulation of the fragments, a prolonged surgical exposure, and might not have yielded the expected results. The end results demonstrate the wisdom of this surgeon's choice.

At the last follow-up, the dog was increasing its use of the limb. Because of the osteopenia, it was suggested that the patient be returned for an additional radiographic evaluation in 3 months. Stress protection is a major factor in treatment of fractures of this type. There is a fine line between providing the proper support of the fragments, which will allow good bone healing, and excessive support, which will result in osteopenia. The plate is to be left in position unless some difficulty becomes evident.

Signalment: 21 months old, female, spayed, Siamese cat

History: cat returned home nonweight bearing on left pelvic limb

Physical Examination: fractured left tibia

Radiographic Examination: left tibia

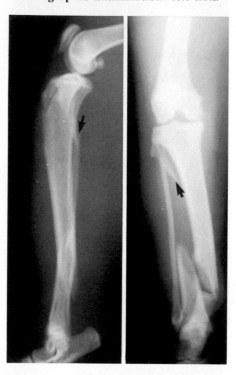

Study at time of entry (day 1):

1. long, oblique fracture of distal tibia
2. lateral displacement of distal fragment with overriding
3. proximal fibular fracture (arrows)
4. radiographically normal stifle and tibiotarsal joints
5. partially open growth plate of tibial crest typical for age

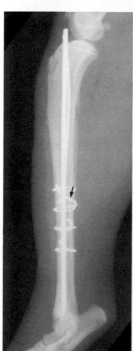
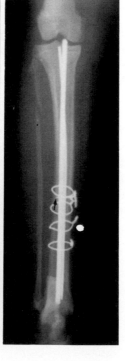

Postoperative study (day 1):

1. anatomic reduction using four full cerclage wires stabilized with two stacked IM pins
2. excellent apposition and alignment of fragments
3. tips of pins extend proximal to tibial crest
4. fibular fracture reduced

Postoperative study (day 42):

1. pins and wires as before
2. fracture lines not visualized
3. minimal external callus identified
4. apposition and alignment of fragments as before
5. healing fibular fracture

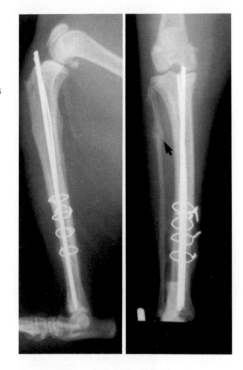

Comments on radiographic findings:

1. the absence of callus in this patient is a consequence of the rigid stability achieved at the fracture site and indicates the lack of requirement for a large external callus; note the comparatively larger callus required to heal the unstable fibular fracture (arrow)

Summary:

In this fracture, the stacked pins prevented rotation of the fragments. Placement of the first pin was followed by placement of the second pin to fill the medullary cavity more completely and provide more rotational control of the fragments. The use of four cerclage wires provided good stabilization of the oblique fracture. The placement of the proximal wire away from the apparent fracture line suggests that a proximal fissure line was noted at the time of surgery. It is important to evaluate for fissure fracture lines because they may not be evident radiographically. The surgeon recognized the slight increase in size of the tibial shaft and ensured that the proximal wire would not slide by placing a small K wire (arrows) in a transfixation technique just distal to the cerclage wire.

The apparent protrusion of the tips of the pins proximally appears on the radiograph to be a surgical error. The tips of the pins are within the intercondylar area, and do not interfere with extension of the limb. The pins could have been shorter, but it is necessary to leave a portion available for grasping at the time of removal. More important, note that the distal point of the IM pin remains within the epiphysis.

The fibular fracture was left untreated. It is interesting to note the degree of callus formation around this unsupported fracture.

Use of stacked pins in an active dog might not have been successful, because it has been shown that double pins are not as strong as previously believed. Also, it is not possible in the dog to seat the pins as solidly in the distal end because of the curvature of the tibia and the narrowing of the medullary cavity. The use of multiple, well placed, stabilized cerclage wires provided excellent support in this fracture, and posed no risk to the developing extraperiosteal blood supply.

CASE 9 Signalment: 5 years old, male, intact, Golden Retriever

History: dog taken to emergency clinic after returning home lame

Physical Examination: fractured left tibia

Radiographic Examination: left tibia

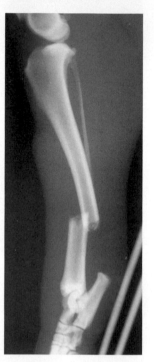

Study at time of entry (single lateral view) (day 1):

1. transverse fracture at junction of middle and distal thirds of tibia
2. minimal comminution at fracture site
3. minimal overriding of fracture fragments
4. radiographically normal stifle and tibiotarsal joints
5. limb immobilized by Schroeder-Thomas splint

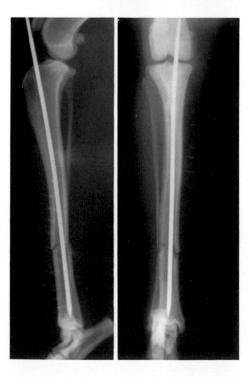

Postoperative study (day 3):

1. reduction using a single IM pin with slight distraction of the fragments
2. good apposition and alignment of fragments
3. tip of the pin has not been cut

Postoperative study (day 41):

1. IM pin remains as before, having been cut proximally
2. gap at fracture site remains
3. callus is forming at the fragment ends, but not bridging
4. apposition and alignment of fragments as before

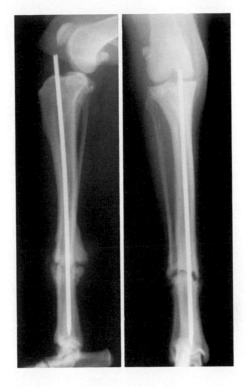

Comments on radiographic findings:

1. the craniocaudal view of the tibia with the splint in position often is of little value because of the overlying radiopaque shadows—additional oblique radiographs can be made with the splint in place, or radiographs can be made at the time of surgery
2. the callus formation on day 41 arises from the ends of the fragments—the density and failure to bridge the fracture site suggest a potential nonunion

Summary:

The surgeon has chosen an IM pin that is too small for stabilization of this transverse fracture without additional support. In a transverse fracture, a single pin permits rotary motion even though it is deeply seated. The pin is not firmly held within the medullary cavity of the distal fragment.

The way in which the callus formed indicates a good blood supply to the bone ends, but continuous motion has disrupted the developing capillary formation at the fracture sites. As a result, heavy "beaks" of new bone are trying to bridge the fracture site but fail to do so. The limb was supported by a Schroeder-Thomas splint during healing, but, as one might expect, it provided little support and, in fact, may well have acted as a pendulum, distracting and moving the fragment.

A simple fracture such as this one would have been ideal to plate using at least an eight-hole plate with four screws proximally and four screws distally. Transverse fractures appear deceptively easy to treat, frequently misleading the surgeon into using simpler methods that, over the long term, disrupt congenial client relationships.

Signalment: 2 years old, male, castrated, domestic short-haired cat

History: referred from emergency clinic after probable hit by car

Physical Examination: open fracture of right tibia, possible pelvic fractures, difficult to evaluate neurologically because of the multiple injuries

Radiographic Examination: right pelvic limb

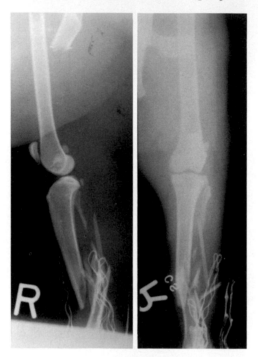

Study at time of entry (day 1):

1. transverse fracture at junction of proximal and middle thirds of femur with minimal comminution
2. marked overriding of fracture fragments
3. oblique comminuted midshaft tibial fracture
4. marked overriding of fracture fragments
5. multiple fractures of the fibula
6. soft tissue injury, creating an open fracture
7. note radiopaque markers within surgical sponge
8. stifle joint appears normal; tibiotarsal joint not completely evaluated

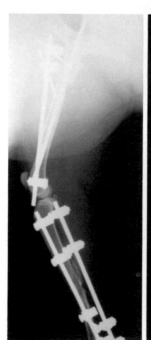
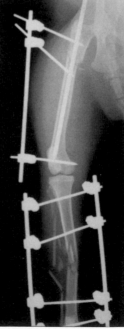

Postoperative study (day 5):

1. reduction and stabilization of femoral fracture using a single IM pin and a modified type I K-E apparatus with one pin proximally and two pins distally
2. good apposition and alignment of the fragments with retroversion of the femoral head and neck
3. reduction and stabilization of tibial fractures using a type II K-E apparatus (2/2)
4. good apposition of fragments with minimal lateral and cranial angulation of the distal fragment

Postoperative study on day of second surgery (tibia only) (day 93):

1. K-E apparatus removed and a ten-hole bone plate applied with four screws proximal and four screws distal
2. no radiographic evidence of cancellous graft
3. fracture site without callus formation
4. apposition and alignment of fragments good with minimal medial angulation of distal fragment
5. fixation pin holes remain
6. question instability of stifle joint

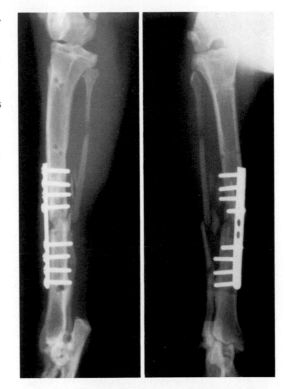

Second postoperative study (tibia only) (day 103):

1. first bone plate removed because of bending
2. second larger and longer eight-hole bone plate applied
3. early healing of fracture with slight malunion
4. no restoration of medullary cavity or cortical shadows
5. good apposition and alignment of fragments, with length preserved
6. marked soft tissue atrophy
7. fixation pin holes remain

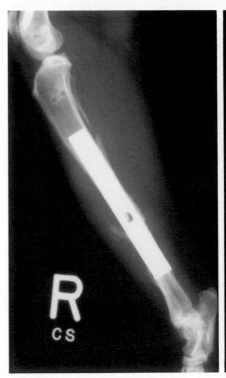
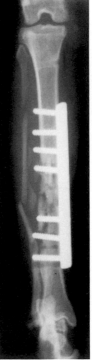

Comments on radiographic findings:

1. avoid including two long bones on the same radiographic study—it is difficult to evaluate completely both femoral and tibial fractures—the hip joint and tibiotarsal joints are incompletely evaluated
2. note the radiopaque markers within the surgical sponges wrapped around the limb; although they are good as markers to search for "lost" sponges, they are not helpful in this radiographic study
3. status of the soft tissues can be evaluated radiographically—note the marked soft tissue atrophy as well as the instability at the stifle joint

Summary:

This patient suffered multiple fractures, with the femoral fracture handled with a single IM pin and a modified type I K-E apparatus providing additional support. The tibia was treated in a seemingly acceptable manner using a type II K-E apparatus. The fixation pins are noted to be unthreaded.

Studies for the next 75 days failed to identify a healing bridging callus. Because of this, the tibial fracture was plated; however, the plate was too light and too short for this size of bone. As is demonstrated clearly, plate selection is essential. They must be long enough and strong enough to provide support for good healing. (This is described on page 48 in the second edition of the handbook by Brinker, Piermattei, and Flo.) The plate selected was obviously too light, and bent 3 weeks after application. A second heavier and longer plate was used that allowed for the fracture to heal several months later.

The original treatment used for the tibial fracture should have worked. However, the external fixation did not provide satisfactory support. The selection of a bone plate was proper, but much too light for an animal as active as this one, which jumped from a height during healing. Fortunately, the owner permitted the placement of a final plate that led to healing.

Signalment: 3 years old, female, spayed, German Shepherd Dog

History: dog found recumbent in morning with severe injury to left pelvic limb

Physical Examination: gunshot wound to left tibia

Radiographic Examination: left tibia

Study at time of entry (day 1):

1. highly comminuted open fracture at junction of middle and distal thirds of tibia with massive bone loss
2. fracture lines extend to the proximal third of the tibia
3. soft tissue defect
4. marked displacement of fracture fragments
5. distal fibular fracture
6. small metallic fragments within soft tissue
7. tibiotarsal joint appears normal

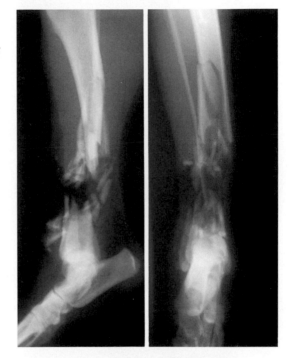

Postoperative study (day 10):

1. stabilization using transfixation pins with one pin positioned proximally and one pin distally in the tibia, and one pin in the calcaneus
2. plaster of Paris cast
3. cranial displacement of the distal fragment with probable shortening of the bone

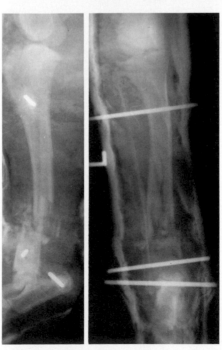

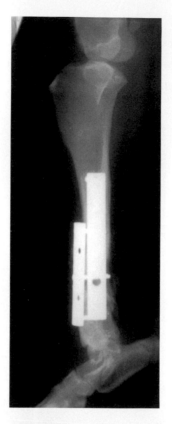

Second postoperative study (single lateral view) (day 53):

1. cast and transfixation pins removed
2. stabilization using two bone plates, with two screws in one plate and five screws in the second plate
3. cancellous graft cranially
4. cranial displacement of the distal fragment with probable shortening of the bone

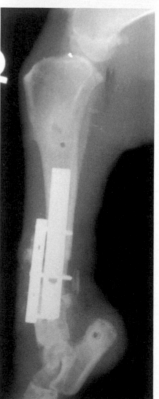

Postoperative study (single lateral view) (day 134):

1. stabilization achieved using two bone plates
2. minimal callus
3. cranial displacement of distal fragment as before with continued shortening of the bone
4. disuse osteopenia

Comments on radiographic findings:

1. radiographic evaluation of the tibiotarsal joint is compromised by the soft tissue gas, and the possibility of an articular component should not be completely excluded
2. the absence of larger metallic fragments suggests very high-energy injury from steel-coated bullet
3. correct length of the bone should be determined from radiographs of the opposite limb
4. early callus formation cannot be evaluated through a plaster cast
5. callus formation in a fracture repaired with two bone plates is difficult to evaluate

Summary:

A very high-energy gunshot wound with massive bone loss and soft tissue injury is an extremely difficult fracture to treat, especially in the region of the distal tibia. Transfixation pins were used to maintain the length of the bone and add stability during the time the soft tissue was being treated. There was space available for the use of more pins. The cast was used to stabilize the pins, but this covered the extensive soft tissue wound and made its treatment difficult. The use of external fixation might have offered a much better opportunity for treatment of the soft tissue injuries and maintenance of fragment stability.

The owner must be informed that treatment using the transfixation pins and a cast may lead to healing, but there is a strong possibility that additional surgery will be required. After almost 2 months, the soft tissues had healed but the bone failed to unite, and the surgeon elected to use two short plates to stabilize the fracture. After 4 months, soft tissue necrosis occurred and the plates became exposed. The shorter plate was removed, a nonunion was found, and the dog was euthanized. This unfortunate outcome was probably due to the severity of the injuries to the soft tissue rather than to those of the bone.

CASE 12 **Signalment:** 12 years old, female, intact, Norwegian Elkhound
History: dog struck by a golf ball
Physical Examination: fractured left tibia
Radiographic Examination: left tibia

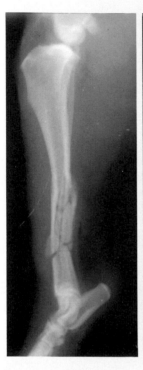
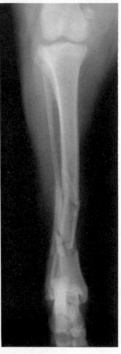

Study at time of entry (day 1):

1. highly comminuted segmental fracture at junction of middle and distal thirds of tibia
2. segmental fragments are split longitudinally
3. minimal displacement of fracture fragments
4. radiographically normal stifle and tibiotarsal joints

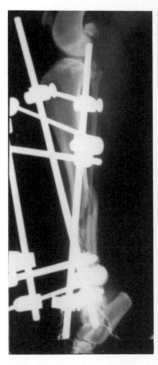
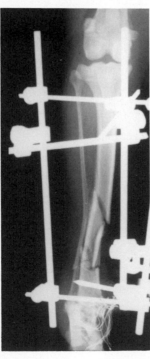

Postoperative study (day 1):

1. reduction using a type III K-E apparatus with three pins proximally and two pins distally
2. good apposition of fragments with lateral displacement of the distal fragment

Postoperative study (day 38):

1. K-E apparatus as before with marked lucency around a proximal fixation pin (osteomyelitis) (arrows)
2. fracture lines remain evident
3. aggressive periosteal new bone formation along the shaft of the bone (osteomyelitis)
4. apposition and alignment of fragments as before with internal rotation of the foot

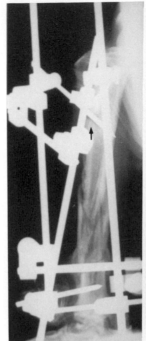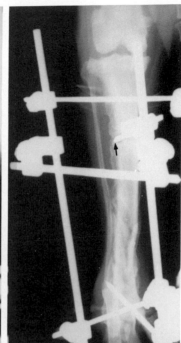

Postoperative study (day 57):

1. K-E apparatus as before with continued pin loosening (osteomyelitis)
2. fracture lines remain evident
3. external callus formation without pattern, but bridging the fracture site
4. apposition and alignment of fragments as before

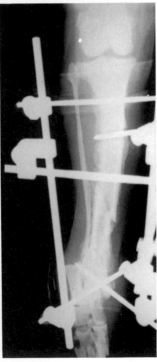

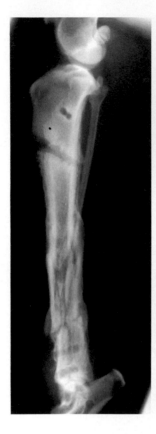 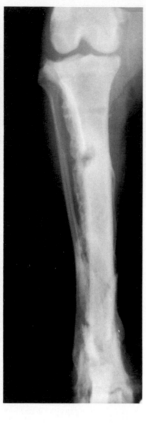

Postoperative study (day 57):

1. removal of K-E apparatus
2. pin holes are larger than expected due to osteolysis associated with the osteomyelitis
3. fracture lines remain evident
4. external callus formation without complete bridging (delayed healing)
5. apposition and alignment of fragments as before

Comments on radiographic findings:

1. the periosteal response on day 38 involves the entire shaft of the tibia; this plus the lucency around a proximal pin supports a diagnosis of osteomyelitis
2. note how much easier it is to judge healing on the last study, after removal of the apparatus

Summary:

Consistent with all fractures treated with external fixators, this reduction is not as anatomic as with a bone plate. However, the exposure is limited through use of a blind approach in placement of fixation pins. A cranial component was placed to create a "tenting" pattern. The use of double clamps permitted the interconnection of the fixation pins. Note that the position of the most distal fixation pin was close to the articular surface.

Later, the fixation pins of the cranial component became loose, as evidenced by the lucent zones around them. Because of the lucent zones around the pins and the aggressive pattern of periosteal new bone, a diagnosis of osteomyelitis is warranted. Fracture healing seems to be progressing, but is delayed.

The surgeon selected this mode of treatment because of the extent of soft tissue damage. The surgeon considered that placement of a bone plate would be most difficult because of the short distal fragment. The tenting technique provides maximum stability, and might have been more quickly successful in this patient but for the onset of osteomyelitis. This form of stabilization is satisfactory; however, a bone plate with interfragmentary compression still might have been used. After removal of the external apparatus, the lytic appearance of the pin tracks and nature of the periosteal response can be observed more easily radiographically. Fracture lines are still evident, with suspected sequestration of the original fracture fragments. The patient, examined 1 month later, showed early clinical healing. The limb was supported for an additional period of time with an external splint.

Although the severe periosteal response gives rise to a diagnosis of osteomyelitis, it may well have been the result of the severe soft tissue injury or soft tissue infection. Radiographic findings always should be supported by the clinical findings. The fact that the patient used the limb, albeit tentatively, throughout the healing period would indicate that the surgeon might appropriately delay aggressive further treatment while observing the progress of the lesion.

CASE 13 Signalment: 3 years old, male, castrated, domestic short-haired cat

History: cat shot by irate neighbor, K-E apparatus applied 2 days before referral

Physical Examination: suspected fracture of right tibia

Radiographic Examination: right tibia

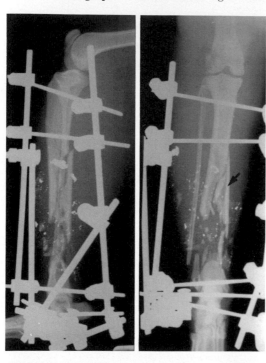

Study at time of entry (day 3):

1. very high-energy fracture of right tibia with severe comminution, a large butterfly fragment (arrow), bone loss, and extensive soft tissue injury without apparent joint injury
2. apposition and alignment achieved by placement of a modified type II K-E apparatus with two fixation pins proximally and three fixation pins distally, two of which are in the tarsal bones
3. distal fixation pin probably enters the tibiotarsal joint
4. radiographically normal stifle joint
5. multiple radiodense metallic fragments secondary to gunshot injury

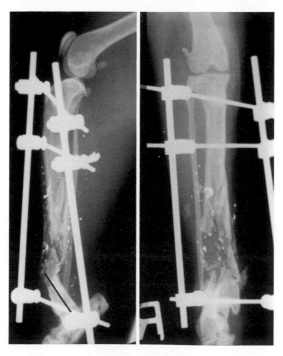

Postoperative study (day 48):

1. removal of distal component of the K-E apparatus
2. apposition and alignment of fragments has shifted with marked caudal angulation of the distal fragment (line)
3. no evidence of callus formation
4. proximal fracture lines remain visible, suggesting delayed healing
5. soft tissue swelling has regressed
6. multiple radiodense metallic fragments as before

Postoperative study (day 75):

1. remainder of K-E apparatus has been removed
2. no evidence of callus formation
3. apposition and alignment of fragments as before, with marked shortening
4. multiple radiodense metallic fragments as before
5. soft tissue atrophy
6. nonunion fibular fracture

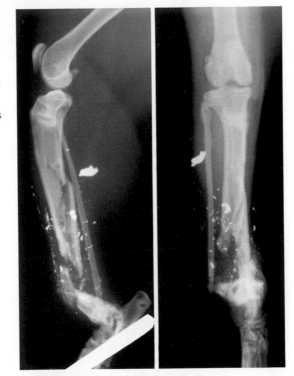

Postoperative study (day 111):

1. reduction of fragments using a bone plate with four screws proximally and two screws distally
2. good alignment of fragments with 2 cm shortening of the tibia
3. fracture lines less distinct, suggesting healing
4. multiple radiodense metallic fragments as before
5. tibiotarsal joint collapse
6. pin holes within the tarsal bones
7. nonunion fibular fracture

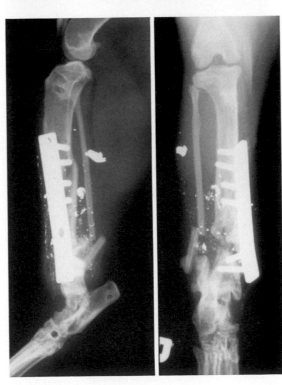

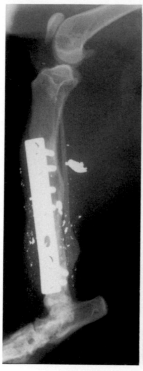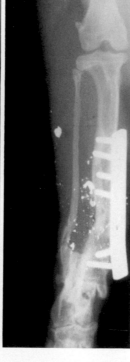

Postoperative study (41 months):

1. plate and screws remain in position with persistent tibial shortening
2. good bridging at fracture site with minimal external callus
3. fracture lines not visualized
4. external callus formation
5. lateral angulation of distal fragment
6. fibular fractures healing
7. pin holes within the tarsal bones are filling

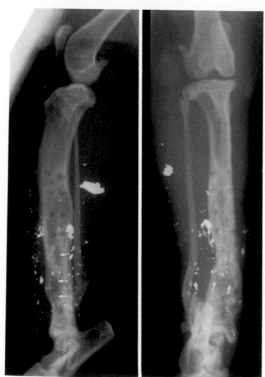

Second Postoperative study (41 months):

1. removal of bone plate and all screws
2. malunion all fractures
3. lateral angulation of distal fragment as before
4. pin holes remain unfilled
5. metallic fragments remain as before
6. posttraumatic joint disease in tibiotarsal joint
7. stifle joint radiographically within normal limits

Comments on radiographic findings:

1. external callus formation is greatly delayed due to soft tissue injury—internal callus eventually formed resulting in bony healing
2. determination of the presence of bone infection during the healing of this fracture is almost impossible radiographically—clinically, the patient had episodes of limb disuse, soft tissue swelling, and soft tissue drainage, all of which suggest infection
3. examination for sequestration on last study showed suspect lucent areas (arrows) within the bone but no evidence of bone sequestra—cortical modeling had not yet taken place
4. the tibiotarsal joint could have been studied independently to determine the progress of joint disease after the intra-articular placement of a fixation pin

Summary:

This is an interesting case in which a gunshot wound caused extensive bone loss and severe soft tissue injury, with only a short distal tibial fragment remaining. The referral surgeon was of the opinion that the distal fragment was not of sufficient length to permit use of a bone plate and, desiring to maintain the length of the tibia, the decision was made to use an external K-E device. A modified type II K-E was applied, using, in addition to the tibia, the calcaneus and middle row of tarsal bones for placement of distal fixation pins. The distal tibial pin did not engage the fragment satisfactorily; not only did it fail to control the fragment, it unfortunately damaged the tibiotarsal joint. It is not clear whether threaded pins might have assisted in the treatment.

After almost 4 months with no signs of union, a bone plate and cancellous graft were applied. To approximate the fragments, tibial length was sacrificed. Only one screw was firmly seated within the distal fragment. An external cast was used in addition to the plate. However, the cast became loose and "fell off" within 1 month after application. At this time, the patient began to use the limb. The plate was removed 41 months later because of soft tissue drainage. At this time, the fracture appeared to be healed and the infectious process was found to be limited to the overlying soft tissues.

Once stability was established with placement of the plate, healing, although slow, continued, and the cat began to use the limb, demonstrating increasing weight bearing. In this nonathletic patient, the shortening of the limb was easily compensated for.

The advantages of fragment stability, in addition to fracture healing, were the increased use of the limb by the patient and freedom from pain. The decision to plate the fracture could have been made much earlier, because it is obvious from the radiographic studies that there was motion at the fracture site with an absence of callus formation.

Tarsus–Metatarsus

Signalment: 10 months old, male, intact, cat

History: cat found near road 24 hours ago with injury to the right pelvic limb

Physical Examination: open wound with laxity of tarsometatarsal joint

Radiographic Examination: right hind foot

Study at time of entry (day 1):

1. fractures of the proximal second, third, and fourth metatarsal bones and luxation of the head of the fifth metatarsal bone (arrow)
2. small butterfly fragment, second metatarsal bone
3. marked angulation of fracture fragments
4. tibiotarsal joint is not parallel with proximal intertarsal joint, suggesting acute crushing injury to talus (lines)
5. status of the tarsometatarsal joints uncertain
6. tibiotarsal joint appears normal
7. physeal closure suggests an age older than that reported by the owner

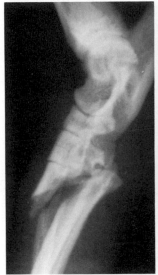
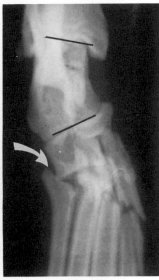

Postoperative study (day 3):

1. reduction using a cerclage wire on the second metatarsal bone and a single cerclage wire around both the third and fourth metatarsal bones
2. a single IM pin was placed through each of the second, third, and fourth metatarsal bones, with the pins seated within the tarsal bones
3. lateral displacement of third, fourth, and fifth metatarsal bones
4. minimal lateral angulation of the distal fragments
5. status of the talus still is not evident on this study

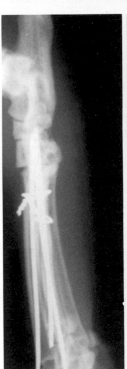
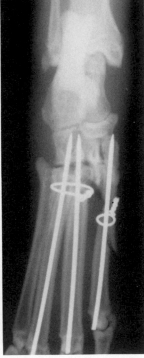

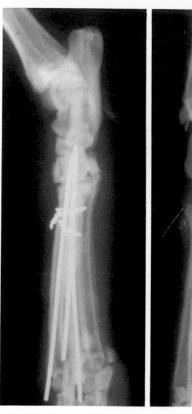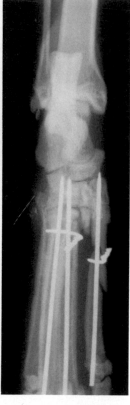

Postoperative study (day 28):

1. persistent malalignment of fragments
2. fracture lines still identified
3. no evidence of callus formation
4. metallic devices as before
5. status of the talus still is not clear

Comments on radiographic findings:

1. the dorsoplantar view on day 1 suggests that tarsal bone damage may be severe, with an apparent crushing injury to the talus
2. radiographic evaluation of the fracture–luxation is difficult because of the small size of the bones and the overriding of the metatarsal bones
3. after reduction, more complete evaluation of the joint injury can be made
4. the ligamentous injury in this patient is severe and difficult to evaluate radiographically

Summary:

Severe damage with ligamentous injury made the reduction of the metatarsal bones difficult. The second metatarsal bone fractures were reduced and stabilized by use of an IM pin and single cerclage wire. The third and fourth metatarsal bones have been stabilized with IM pins but are located laterally and not anatomically reduced. The fifth metatarsal bone remains luxated laterally.

The IM pins enter the distal metatarsal bones dorsally to avoid the joint space. The pins are secured deeply in the tarsal bones, although the pin in the third metatarsal bone is misguided. Customarily the distal ends of the IM pins are bent dorsally at a right angle. This provides a "handle" for removal and avoids injury to the metatarsophalangeal joint.

A radiographic study made 3 months postinjury showed healing of the metatarsal fractures; however, the tarsometatarsal luxation involving the fifth digit remained. The cat had no evidence of any problems with ambulation.

Signalment: 7 months old, female, intact, German Shepherd Dog

History: struck by a car 24 hours ago; not walking

Physical Examination: nonweight bearing on right pelvic limb, paw is deviated medially, marked soft tissue swelling of foot

Radiographic Examination: left pelvic foot (dorsoplantar and oblique views only)

Study at time of entry (single dorsoplantar view) (day 1):

1. oblique, slightly comminuted midshaft fractures of the second, third, and fourth metatarsal bones, with a degree of displacement and medial angulation of the distal fragments
2. greenstick fracture at junction of middle and distal thirds of the fifth metatarsal bone with a persistent cortical shadow with fragments in near-alignment (arrow)
3. bandage material casts a greater-density shadow around the toes

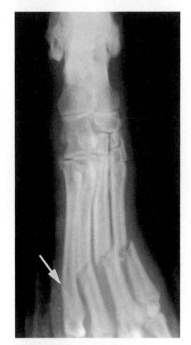

Postreduction study (single oblique view) (day 3):

1. obliquely placed K wires cross the fracture sites of the second and third metatarsal bones
2. distal fracture fragments remain slightly angled medially

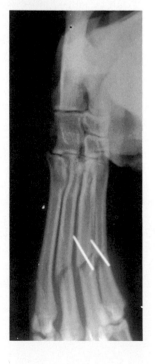

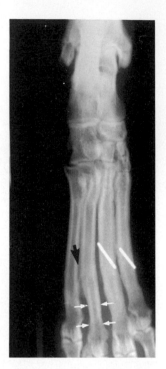

Postreduction study (single dorsoplantar view) (day 30):

1. wires in the second and third metatarsal bones remain in position
2. distal fracture fragments remain slightly angled medially
3. callus forming at all fracture sites
4. fracture lines only faintly visible
5. "penciling" of distal metatarsal bones (white arrows)
6. question as to an early synostosis between fourth and fifth metatarsal bones (black arrow)

Comments on radiographic findings:

1. a fracture with an intact cortical shadow must be an incomplete fracture
2. multiple radiographic views are required to evaluate injuries to the foot
3. "penciling" of the distal metatarsal bones as seen on the last study is common and is probably due to disuse (white arrows)
4. cross healing between the fourth and fifth metatarsal bones (black arrow) is suspected but can be judged more accurately on evaluation of all views.

Summary:

Fractures of this type could be treated by placing the foot in a cast or splint. However, all casts must be observed carefully for: (1) reduction of swelling, which causes loosening, (2) pressure sores, and (3) soiling, permitting the cast to become soft and ineffective. The frequent replacement of the cast for these reasons makes for poor immobilization of the bone fragments, and a greater expense than if the fractures had been treated more aggressively at the onset. Internal stabilization is thus considered a better method of treatment.

In this patient, the fragments of the second and third metatarsal bones were reduced. The fragments of the fourth metatarsal bone tended to be held in a more normal position with stabilization of the second and third metatarsal bones. The fifth metatarsal bone will heal well. The K wires in this patient provided alignment and reduction. A cast was used for additional support.

Another method of treatment would be the introduction of IM pins distally to traverse the fracture and be seated proximally. The healing in this patient is considered fortunate because little stabilization of the fragments was achieved by the K wires; however, they did maintain alignment while the foot was in the cast.

Head

Signalment: 10 years old, male, intact, domestic short-haired cat

History: cat unable to eat; some type of trauma suspected

Physical Examination: suspected injury to mandible

Radiographic Examination: head

Study at time of entry (single dorsoventral view) (day 1):

1. fracture line through mandible (left) just rostral to mandibular condyle (arrow)
2. minimal displacement of mandible to the left
3. opposite temporomandibular joint (right) is normal on comparison

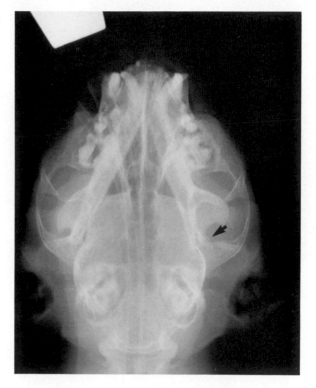

Postoperative study (complete study) (day 17):

1. reduction of the fracture using a single-wire suture
2. good apposition and alignment of fragments

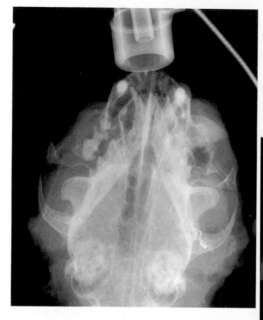

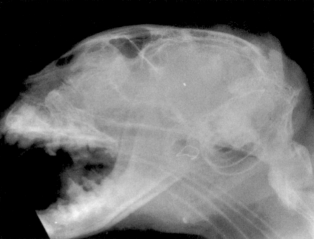

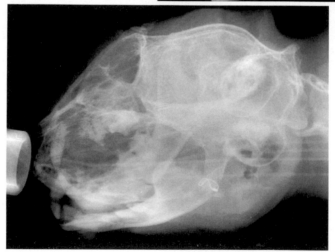

Comments on radiographic findings:

1. mandibular fractures in the region of the temporomandibular joint are difficult to identify—note that after the injury to the left temporomandibular joint malocclusion of the mandible occurred, causing a shift of the mandible to the affected side—malocclusion of the canine teeth is a good indicator of this type injury—the extent of articular involvement is difficult to evaluate from the original radiograph; this is better determined at the time of surgery
2. note the reduction of the mandibular condylar fracture as seen on the postoperative dorsoventral view
3. the oblique views show that the reduction was nearly anatomic
4. callus formation will be minimal and not easily identified radiographically— progress of fracture healing is better measured by physical examination and noting the patient's willingness to use the joint

Summary:

The fracture line through the mandibular notch separates the mandibular condyle (condyloid process or articular process). This is a clinically important injury, and reattachment of some type will enable the cat to eat more easily. The "notch" is "neck-like," and its fracture frequently permits rotation of the condyle. Malposition of the fragment makes its radiographic appearance difficult to appreciate.

The surgeon wired the condyloid process back to the ramus. In small patients, only one wire can be positioned. In the larger dog, a muzzle often is used to hold the mandible in position while awaiting healing. This is possible because the condyle usually remains in a more anatomic position after the fracture.

Nonunion is always a problem when solid reduction cannot be obtained. Were this to happen, the condyle could be subsequently removed, permitting the formation of a false joint.

Signalment: 4 years old, male, intact, mixed-breed terrier

History: dog kicked by horse 2 days ago

Physical Examination: suspected mandibular injury

Radiographic Examination: mandible

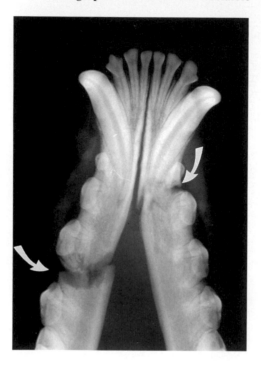

Study at time of entry (open-mouth view) (day 1):

1. oblique fracture of the mandible between the right third and fourth premolar teeth (arrow)
2. oblique fracture of the mandible between the left first and second premolar teeth (arrow)
3. fracture lines enter the periodontal spaces
4. minimal displacement of the rostral fragment to the left
5. remaining periodontal space radiographically normal

Postoperative study (open-mouth and oblique views) (day 30):

1. reduction using multiple-wire sutures embedded within intraoral methylmethacrylate plastic mold
2. good apposition and alignment of fragments
3. reactive bone forming around the fracture sites without bridging (white arrowheads)
4. marked widening of the periodontal space on several teeth (osteomyelitis)
5. suspected sequestra formation (black arrowheads)
6. wire suture around the teeth has broken

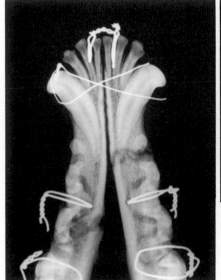 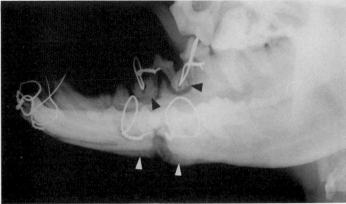

Postoperative study (open-mouth, lateral, and oblique views) (day 44):

1. all metallic hardware has been removed
2. fracture line still visualized, with minimal callus on right
3. fracture line healed on left with bridging callus
4. good apposition and alignment of fragments as before
5. radiolucent pattern around teeth as before (osteomyelitis)
6. two large sequestra on left and one on the right (arrowheads)

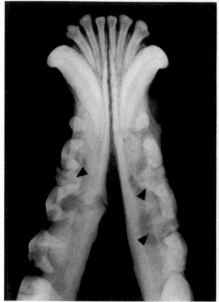
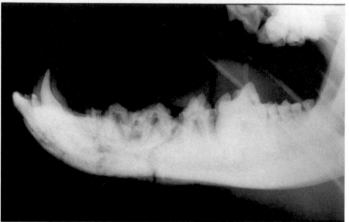
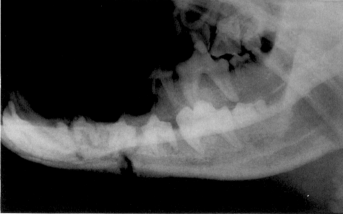

Comments on radiographic findings:

1. fracture lines enter radiographically normal periodontal spaces at the time of original injury—the fractures thus are open both to the oral cavity and to the periodontal spaces; the possibility of osteomyelitis is therefore obvious
2. the methylmethacrylate mold is radiolucent and not visualized on the radiograph
3. temporomandibular joints were judged normal radiographically
4. other views at the time of entry suggested fractures of the crown of both right and left upper fourth premolar teeth
5. development of osteomyelitis is noted in the periodontal spaces
6. sequestra formed at the sites of wire attachment

Summary:

The surgeon used wires to hold a methylmethacrylate mold as a method to stabilize the fracture fragments. A number of drill holes were made in the periodontal tissues to achieve fixation of the wires. The plastic impression mold fit tightly against the teeth and gums, and provided an opportunity for periodontal pocketing of bacteria. The wires provided tracts for the development of active infectious periodontitis (osteomyelitis with sequestration). Periodontitis is a frequent complication of the use of intraoral plastic molds.

Other choices for treatment might have included use of a small bone plate or small transfixation pins. It is possible to position a pin just rostral to the canine teeth. Because there was minimal malpositioning of the fragments, it might have been possible to apply a muzzle during healing.

The treatment was complicated by the presence of osteomyelitis with two rather large sequestra. Curettage is required to remove the infected sequestra and infected bone tissue lining the periodontal pockets. The third premolar on the left appears to have no bony attachment (arrow), and probably will have to be extracted. The use of a plate with a cancellous bone graft would be a suggested treatment.

Pelvis

Signalment: 1 year old, male, intact, mixed-breed Australian Shepherd Dog

History: dog hit by car this morning

Physical Examination: dog was depressed and unwilling to use left pelvic limb

Radiographic Examination: pelvis

Study at time of entry (ventrodorsal view) (day 1):

1. left acetabular fractures
2. medial displacement with impaction of acetabular fragments
3. left sacroiliac separation with cranial iliac displacement
4. sacral fracture of articular process on left
5. pubic symphyseal fracture
6. left ischial fracture caudally (arrow)
7. character of left femoral head and neck uncertain radiographically
8. right hip joint appears normal radiographically
9. metallic pellets from shotgun injury (unrelated)

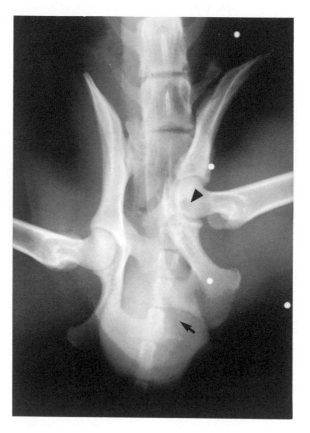

303

Postoperative study (day 2):

1. placement of bone screws and a radiolucent Lubra plate achieve incomplete reduction of acetabular fracture
2. single screw and pin pass through the ilium and fail to enter the sacrum (curved arrow)
3. tension-band device used to reposition the trochanteric osteotomy
4. sacral fragment left *in situ* (arrowhead)
5. metallic pellets as before
6. soft tissue gas

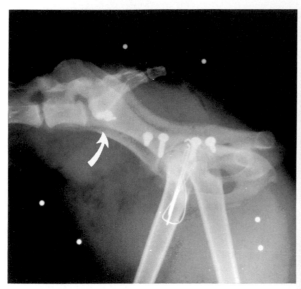
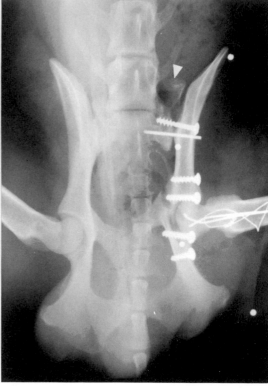

Postoperative study (ventrodorsal view) (day 66):

1. acetabular screws remain in position
2. slight cranial shifting of left hemi-pelvis
3. no radiographic evidence of callus formation around the acetabular fractures
4. pubic and ischial fractures as before
5. tension-band device as before
6. metallic pellets as before

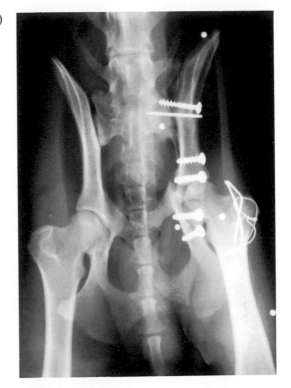

Postoperative study (ventrodorsal view) (day 235):

1. acetabular screws remain in position
2. displaced left hemipelvis as before, with bony bridging
3. acetabular fractures have bridged
4. sacral fragment as before
5. tension-band device as before
6. metallic pellets as before

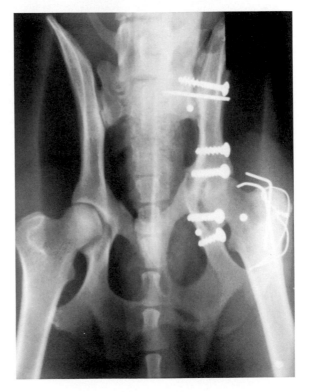

Comments on radiographic findings:

1. fracture lines and/or soft tissue shadows and/or soft tissue gas have created lucent shadows crossing the left femoral head (arrowhead)—in this patient, they are due to the dorsal acetabular fractures and are not the result of injury within the femoral head or neck—a "frog-leg" view changes the positioning of the femoral head and was helpful in solving this problem—if the fracture lines remain the same, they are in the acetabulum; if the fracture lines change in location or appearance they may be within the femoral head
2. both lateral and ventrodorsal views are needed to determine accurately the location of the sacroiliac screw and pin
3. both lateral and ventrodorsal views are needed to evaluate reduction of the acetabular fragments
4. additional lateral views of the caudal abdomen need to be made in patients with pelvic injuries to assist in the evaluation of urinary bladder status

Summary:

With the fracture line entering the hip joint and inward pelvic collapse, this injury cannot be left untreated. The patient would experience great pain on ambulation and would have difficulty in defecation or, in the intact female, parturition. The fractures must be attended to.

An attempt to reduce the separation of the sacroiliac joint ended in failure, with both screw and wire passing ventral to the body of the sacrum. The surgeon should have used anatomic landmarks to ensure the entrance of the wire and screw into the body of the sacrum.

The acetabular fracture was reduced through placement of a Lubra-plate dorsally. Failure to achieve good reduction is noted. The Lubra-plate is a form of tension band that brings fragments together; however, it permits rotation with subsequent separation. K wires or small-diameter IM pins can be inserted to cross dorsal to the acetabulum to control rotation of the fragments. The two caudal screws are positioned near the location of the ischiatic nerve, and care must be taken to protect that structure. Reduction could have been better. The possibility exists of secondary joint disease affecting use of limb.

The tension-band device was used in a unique manner. The drill hole for the wire passes through the medial cortex instead of the more commonly used lateral cortex. Also, the pins were left with a long bend proximally.

The patient was improved after this repair, but not as well as might have been possible.

Signalment: 1 year old, male, intact, domestic long-haired cat

History: owner found cat last night with painful and weak pelvic limbs

Physical Examination: left pelvic limb swollen, left femoral head appeared to be luxated, lacerations over right tarsus, patient was dyspneic

Radiographic Examination: pelvis

Study at time of entry (day 1):

1. right acetabular fracture with impaction of the femoral head
2. luxation of left femoral head

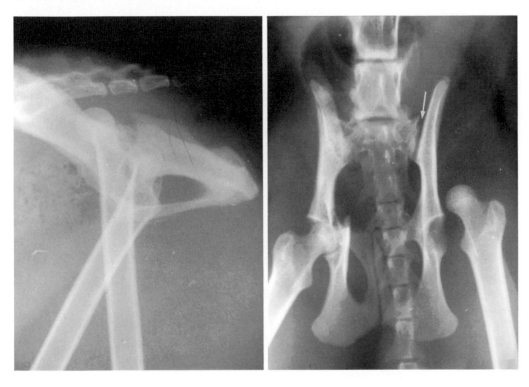

Postoperative study (day 2):

1. placement of bone screws and a single wire achieve incomplete reduction of right acetabular fracture (arrow)
2. left femoral head and neck ostectomy

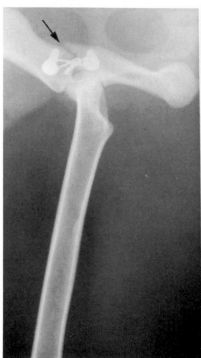 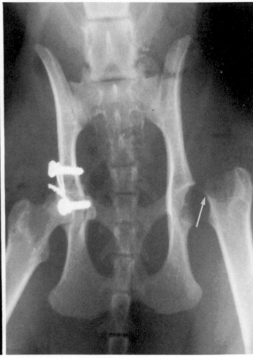

Comments on radiographic findings:

1. left femoral head is located dorsal and lateral to the acetabulum, indicating that soft tissue injury is severe enough to permit this position; with lesser soft tissue injury, the femoral head rides tightly against the ilium just dorsal and cranial to the acetabulum
2. the "open" sacroiliac joint on the left is not traumatic, but due only to oblique positioning (arrow)
3. failure to identify any avulsion fragment within the left acetabulum
4. radiographs of the thorax are important in trauma; in this patient, increased density in the right caudal lung lobe was due to hemorrhage and edema, which explained the patient's dyspnea

Summary:

Surgical reduction apparently was not considered for the luxation of the left femoral head. The choice of femoral head ostectomy is unusual because it is not often considered in the cat owing to the ease of providing a prosthetic joint capsule. The ostectomy was performed with the angle of cut too "flat," leaving the calcar prominent (arrow) where it can come into contact with the acetabulum, a frequent cause of pain.

The right acetabular fracture, if not repaired, would have led to secondary joint disease; it is important to aim for congruent joint surfaces. Two bone screws were introduced on either side of the fracture and a figure-8 tension wire was placed around them to achieve compression. It would have been beneficial to insert a small K wire through the bone fragments in the dorsal acetabular roof to help to maintain alignment before application of the wire. The use of this combination, the figure-8 wire and the pin, is a more effective treatment.

Healing will occur in this patient. The ostectomy may not be as effective as intended because of contact between the calcar and the acetabulum. Ostectomy should be reserved as a salvage procedure, and not used when other methods exist that would preserve the joint.

CASE 3 **Signalment:** 2 years old, male, intact, Brittany

History: dog missing from home for 2 days

Physical Examination: dog was depressed and unwilling to use left pelvic limb

Radiographic Examination: pelvis

Study at time of entry (day 1):

1. left acetabular fracture
2. medial displacement of left femoral head with entrapment
3. left ischial fractures
4. large, elevated "spear-like" fragment dorsal to the acetabulum (arrow)
5. left femoral head and neck normal radiographically
6. right hip joint normal radiographically
7. both sacroiliac joints normal radiographically
8. obstipation

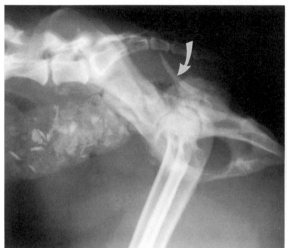
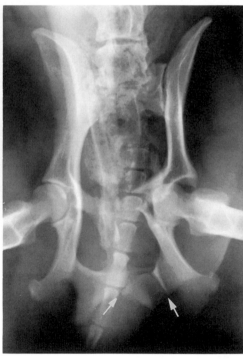

Postoperative study (day 2):

1. placement of single bone plate to reduce left acetabular fracture
2. gap between the ilial fragments (curved arrow)
3. pubic and ischial fragments repositioned

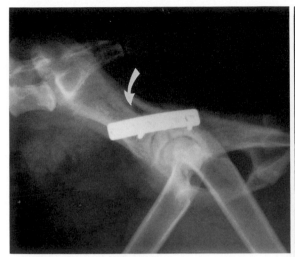
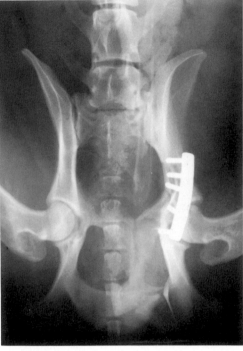

Postoperative study (day 36):

1. bone plate remains in position
2. no radiographic evidence of callus formation around the left acetabular fractures
3. gap between fragments remains as before (curved arrow)

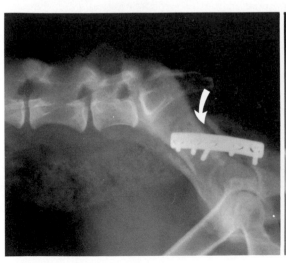
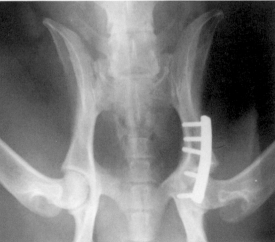

Comments on radiographic findings:

1. obstipation is evidenced by the rectal contents; defecation obviously is painful
2. the character of the hip joints needs to be evaluated carefully; in this patient, there is no problem of injury of the femoral head on the left
3. with most fractures, the radiographic appearance is that of a radiolucent line between fragments—in fractures of flat bones, the fragments often overlie each other, and the radiographic appearance is one of increased density—note the caudal ischial fracture (arrow)
4. the character of the pubic symphyseal fractures often is difficult to determine from pelvic radiographs because of the overlying shadows of the coccygeal vertebral and rectal contents; usually these fractures are not treated, and the absence of absolutely accurate information concerning them is not critical
5. note the use of the "frog-leg" positioning in patients that would experience pain with full extension of the pelvic limbs—position both pelvic limbs in the same manner so that a valid comparison can be made with the normal limb

Summary:

With an acetabular fracture, there is an essential requirement for repair. In this patient, the femoral head is not fractured; however, the imploded femoral head needs to be reduced to avoid a severe, deforming, posttraumatic arthrosis. The surgeon elected to use a bone plate. The surgical approach did not require a trochanteric osteotomy. The plate is positioned without achieving complete fracture reduction. Exposure may have been limited.

The use of a lag screw, or even a K wire or a bone clamp, to hold the fragments before placement of the plate would have achieved a better reduction. Positioning of a plate to reduce a fracture of this type is always difficult because of the craniocaudal bowing and the mediolaterally "rounded" shape of the bone forming the acetabulum. Contouring the plate craniocaudally is relatively easy, but the mediolateral shaping is difficult. For this reason, use of the reconstruction plate has been recommended, although it is a less strong plate. Acetabular plates are produced with both a craniocaudal as well as a mediolateral curvature. The principle is good, but there is such a variation in the sizes required that it is difficult to obtain a plate to fit every requirement. Modification of this type of plate is difficult.

The fracture healed with a minimal amount of secondary joint disease. The dog was clinically sound.

Signalment: 9 months old, female, intact, mixed-breed Golden Retriever

History: unknown history

Physical Examination: dog was depressed and unwilling to use left pelvic limb

Radiographic Examination: pelvis

Study at time of entry (ventrodorsal view) (day 1):

1. ischial fractures in the acetabular branch caudal to acetabulum and in the body near the symphysis
2. pubic fracture with free fragment
3. left sacroiliac separation
4. sacral fracture (large arrows)
5. left hip joint appears normal radiographically except for medial displacement
6. pubic fracture on the right that extends toward the acetabulum (small arrows)
7. cranial displacement of the left ilium

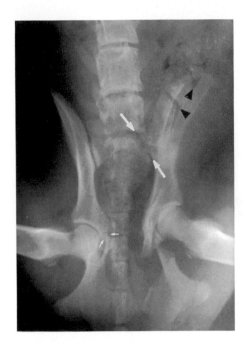

Postoperative study (day 3):

1. placement of single IM pin to reduce the ischial fracture
2. sacroiliac luxation, sacral fractures, and pubic fractures left untreated

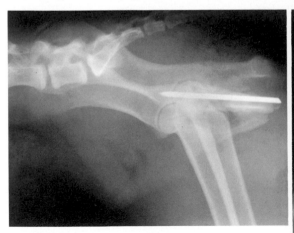
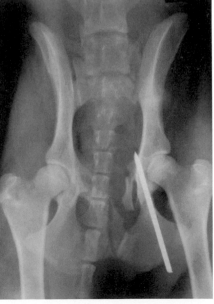

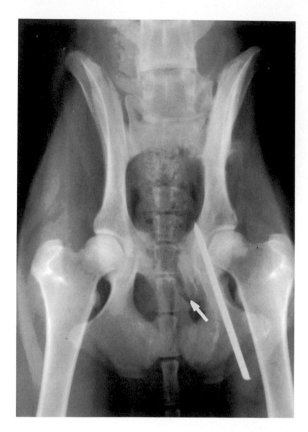

Postoperative study (ventrodorsal view) (day 30):

1. healing of the ischial fractures caudal to acetabulum with heavy callus formation
2. ischial pin remains in position
3. pubic fracture lines still evident (arrow)
4. sacroiliac separation still evident
5. sacral fracture lines difficult to visualize

Comments on radiographic findings:

1. pelvic fractures in a young dog heal with an exuberant callus, especially if perfect anatomic reduction is not achieved; this results in production of an "ugly" pattern of callus formation that sometimes incorrectly suggests bone infection
2. it is important with injuries of this type to evaluate the character of the femoral head and neck, looking for physeal fractures
3. the sacral fracture in this case requires neurologic evaluation
4. note how the gas within the rectum and colon creates the appearance of iliac fracture lines (black arrowheads)

Summary:

With multiple pelvic injuries, a decision needs to be made as to which fracture–luxations require reduction and stabilization. Stabilization of the major ischial fragment is important to relieve pain and reestablish the width of the pelvic inlet. All of the fractures would heal rather quickly, but the repositioning is important to avoid narrowing of the pelvic inlet. The good news in this injury is that the acetabulae and femoral heads and necks are not affected.

Pin placement within the ischium is done through an open reduction to expose the acetabular portion and ensure its central location. The pin is driven from this location caudally until the tip extends from the distal ischium. The protruding pin can be used as a handle to lever the ischial fragment into position. Under direct observation, the fragments are repositioned. The pin is then driven into the cranial fragment of the ischium. Open reduction allows placement of the pin and avoidance of the ischiatic nerve. It is impossible to drive the pin cranially into the shaft of the ilium.

The remaining fragments will be brought into better positioning after reduction of the major ischial fracture. Patient activity may cause the pin to move; to avoid this, it is advisable to use a screw-tipped pin for a better grip on the cranial fragment.

The only way to reposition the left ilium to reduce the sacroiliac luxation is to force it caudally with bone forceps placed on the iliac crest. If the hemipelvis remains intact, it is possible surgically to isolate the ischial tuberosity, and with a bone forceps retract the hemipelvis caudally. After repositioning, the sacroiliac joint is stabilized by use of pins and screws placed through the wing of the ilium into the body of the sacrum. In this patient, the displacement was minimal, so it was left untreated. Also, the associated sacral fracture complicated screw placement. Because of the minimal displacement, the surgeon wisely elected to leave it alone.

Healing occurred quickly, especially in this young dog. The ischial pin needs to remain no longer than 3 or 4 weeks. It should be removed after healing because the dog otherwise cannot sit comfortably.

Index

Note: Page numbers in *italics* indicate illustrations; those followed by t refer to tables.

ISBN 0-7216-5455-X

90038

9 780721 654553